Frontiers in Particle Science and Technology 2014

Particle Interactions Applied

Oak Brook, Illinois, USA
29 April – 1 May 2014

ISBN: 978-1-5108-1761-6

Printed from e-media with permission by:

Curran Associates, Inc.
57 Morehouse Lane
Red Hook, NY 12571

CURRAN ASSOCIATES INC. proceedings.com

Some format issues inherent in the e-media version may also appear in this print version.

Printed by Curran Associates, Inc. (2015)

For permission requests, please contact AIChE
at the address below.

AIChE
120 Wall Street, FL 23
New York, NY 10005-4020

Phone: (800) 242-4363
Fax: (203) 775-5177

www.aiche.org

Additional copies of this publication are available from:

Curran Associates, Inc.
57 Morehouse Lane
Red Hook, NY 12571 USA
Phone: 845-758-0400
Fax: 845-758-2634
Email: curran@proceedings.com
Web: www.proceedings.com

TABLE OF CONTENTS

ACADEMIC KEYNOTE

PACKING OF GRANULAR AND GRANULAR FLUID SYSTEMS

AGGLOMERATION, CLUSTERING AND ATTRITION

BEST PRACTICES IN PARTICLE SCIENCE AND TECHNOLOGY

APPLICATIONS OF NANOPARTICLE TECHNOLOGY

POSTER SESSION

INDUSTRY KEYNOTE

PARTICLE COHESION I

PARTICLE COHESION II

SCALE UP CHALLENGES & SOLUTIONS

The Shapes of Things to Come

Glotzer, S. C., University of Michigan

Originally presented on: 4/30/2014 8:30:00 - 9:00:00

Recent advances in computing make possible the rapid screening, prediction and design of particle shapes optimized for a variety of properties and behaviors. In this talk we present new results on particle packings and assemblies and show how dense and crowded structures depend upon features in "shape space". We provide guidelines and new approaches to particle shape design with photonics, pharmaceutical, biomedical, energy, packaging, cryptography, and many other potential applications.

Assessing the Impact of Particle Breakage on Bed Packing Behavior

Tiemens, B., UOP, A Honeywell Company

Originally presented on: 4/30/2014 9:10:00 - 9:35:00

Many commercial processes in the refining and petrochemical industry utilize packed beds of catalyst or adsorbent particles. When these particles break, the nature in which they pack together can change and negatively impact the performance of the packed bed. Predicting these effects using standard quality control methods for particle mechanical strength is difficult given the complex nature of packed assemblies of irregularly shaped particles. This talk will present several examples of how advanced characterization tools in conjunction with controlled bed packing studies have been used to study the impact of particle breakage in packed beds. High speed video imaging of rapid breakage events reveal different fracture mechanisms of UOP catalysts and how these effects cause differences in quality control tests. Additionally, 3D x-ray tomography (XRT) has been successfully used to non-destructively characterize both individual catalyst particles and sub-scale packed catalyst beds both before and after particle breakage has occurred. Findings from these studies can be used to develop a fundamental understanding of the impact of catalyst breakage on commercial operations.

ITASCA™

MULTIPHYSICS SIMULATIONS OF RIGID AND DEFORMABLE DISCRETE BODIES

Matthew Purvance
mpurvance@itascacg.com

Frontiers in Particle Science & Technology

www.itascacg.com 1 May 2014

Outline

- Brief Introduction to the Discrete Element Method (DEM)
- DEM Modelling of Carbon Nanotubes (CNTs)
- CNT Characterization
- DEM Modelling of Magnetic Dipoles
- Self-Assembly Scoping Study
- Induced Magnetization Scoping Study

DEM with *PFC 5.0* Parlance

- Calculate finite displacements and rotations of discrete bodies

Bodies: ball clump wall

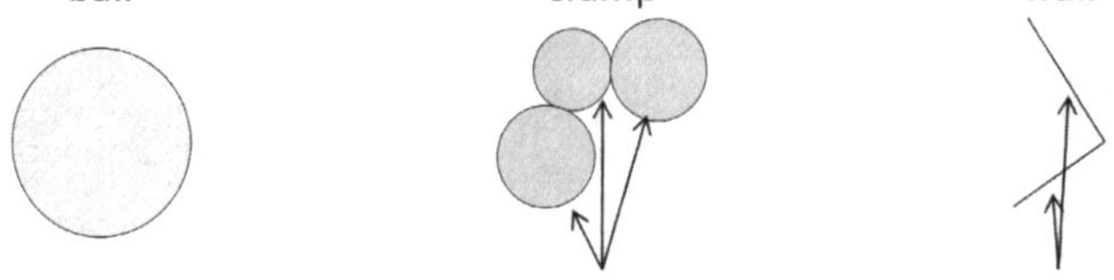

Pieces: ball pebble facet

- Recognize proximity of bodies, creating new contacts (modeled as soft contacts) automatically between pieces while guaranteeing contacts exist prior to force

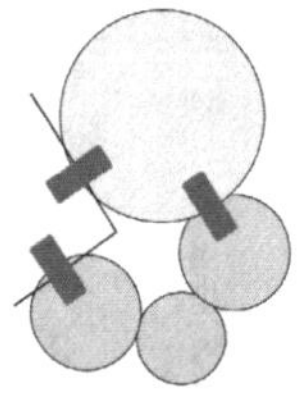

- Explicit time-stepping algorithm including robust time step calculation

Outline

- Brief Introduction to the Discrete Element Method (DEM)
- DEM Modelling of Carbon Nanotubes (CNTs)

- CNT Characterization
- DEM Modelling of Magnetic Dipoles
- Self-Assembly Scoping Study
- Induced Magnetization Scoping Study

Why Study CNT's?

- **Carbon nanotubes (CNTs)** are artificially synthesized carbon structures, possessing exceptional physical and mechanical properties.

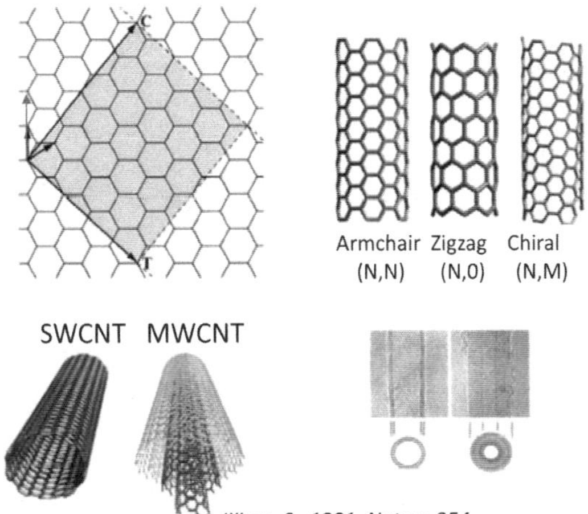

Iijima, S., 1991. Nature 354.

- Young's modulus ~1 TPa
- Tensile strength ~ 100-200 MPa (elongation before yield ~ 10-20%)
- High aspect ratios (1 nm X 20 cm)
- High thermal conductivity, high surface-to volume ratio, tunable electrical properties.

CNT Materials

- Pristine CNT materials - disordered (buckypapers) or highly aligned (CNT ropes and CNT forests) assemblies of CNTs.
- CNT composites – CNT/polymer films, CNT/ceramic or CNT/metal composites.

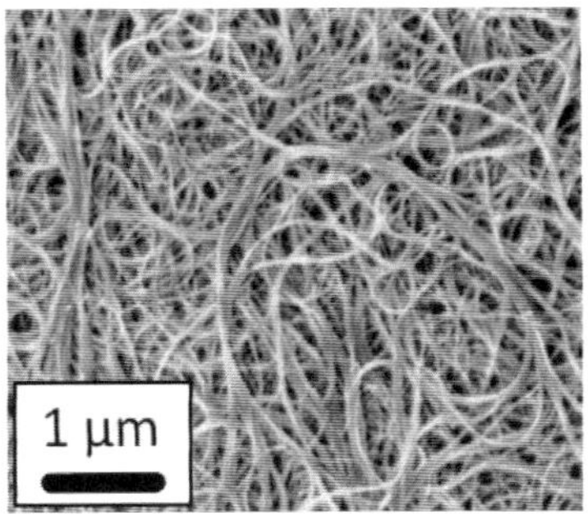

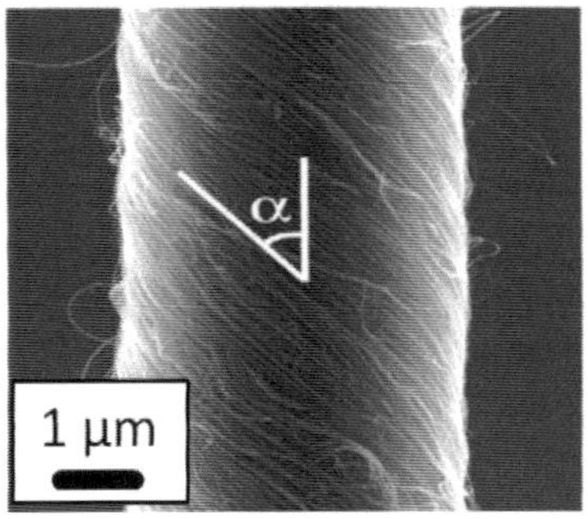

Mesoscopic CNT Models

Mesoscopic models are bridging the atomic-level simulations and continuum level macroscopic properties; they allow to study the mechanical behavior of thousands of CNTs on a time scale of 1÷100 ns.

- **Bead and spring model**
  (Buehler, M., 2006, Journal of materials research, 21(11).)
- **Mesoscopic force field model**
  (Zhigilei, L., et al., 2005, Physical Review B, 71(16).)
- **Distinct element model**
  (Ostanin, I., et al., 2012, Journal of the Mechanics and Physics of Solids, 61(3).)

CNT Specific Contact Models

DEM is supplemented with two additional contact models, to represent **covalent bon[ding]** and **weak vdW bonding**, both contact models create **forces and moments**

1. Mass and stiffness properties of a CNT

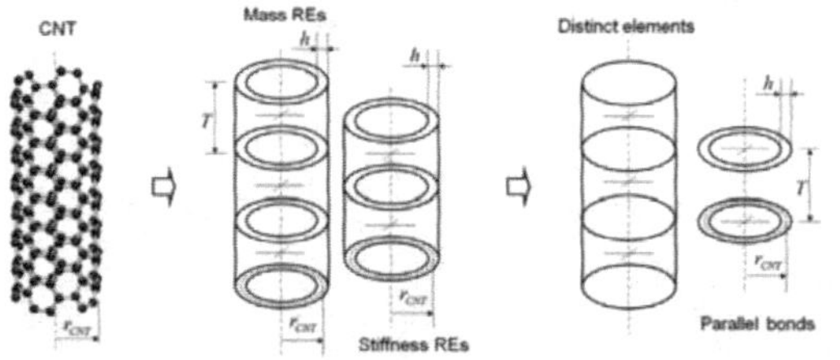

2. van der Waals interactions between CNT

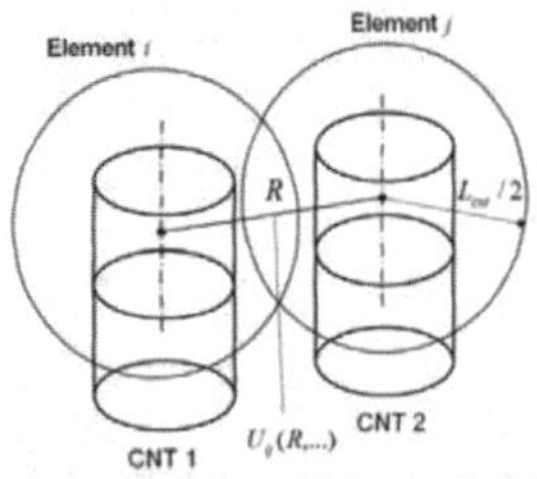

Contact is detected if $R<L_{CNT}$

Potential of interaction depends on mutual orientation between CNT segments

Parallel-bond Contact Model

Representative element of a CNT is viewed as an elastic shell

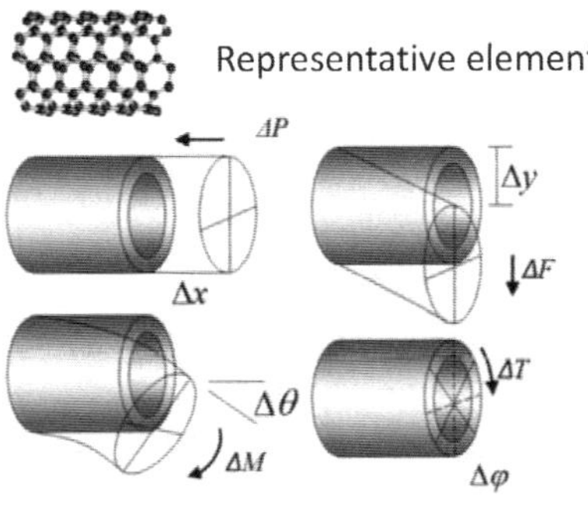

$$\Delta P = -k_n A \cdot \Delta x$$
$$\Delta F = -k_s A \cdot \Delta y$$
$$\Delta M = -k_n I \cdot \Delta \theta$$
$$\Delta T = -k_s J \cdot \Delta \varphi$$

$$k_n = E / L$$
$$k_s = G / L$$

A – cross section area
I – cross section moment of inertia
J – cross section polar moment of inertia

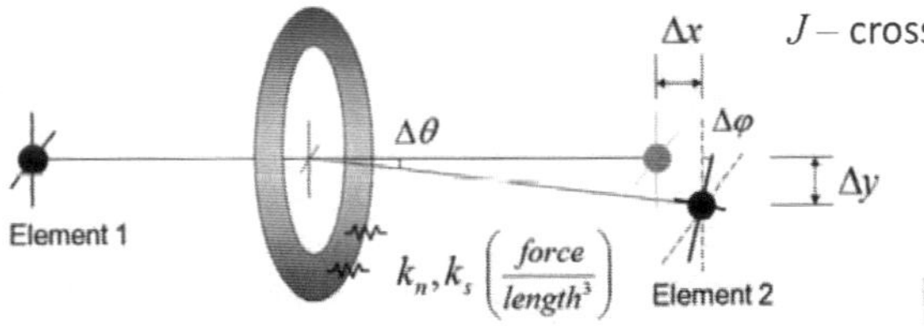

$k_n, k_s \left(\frac{force}{length^3}\right)$

$E = 1.029\ TPa$, $G - 0.459\ TPa$ (determined from atomic-scale simulations)

Potyondy, D., Cundall, P., 2004, Int. J. Rock Mech. & Min. Sci., 41(8).

Van der Waals Contact Model

The contact model for vdW interaction is based on microscopic integration of LJ potential for parallel tubes

$$V = \rho^2 \int_{S_1} \int_{S_2} u(d) dA_1 dA_2$$
$$u(d) = 4\varepsilon \left(\left(\frac{\sigma}{r}\right)^{12} - \left(\frac{\sigma}{r}\right)^{6} \right)$$
$$V(D)/T = 4\varepsilon' \left(\frac{A}{D^\alpha} - \frac{B}{D^\beta} \right)$$
$$D = L / r_{CNT} - 2$$

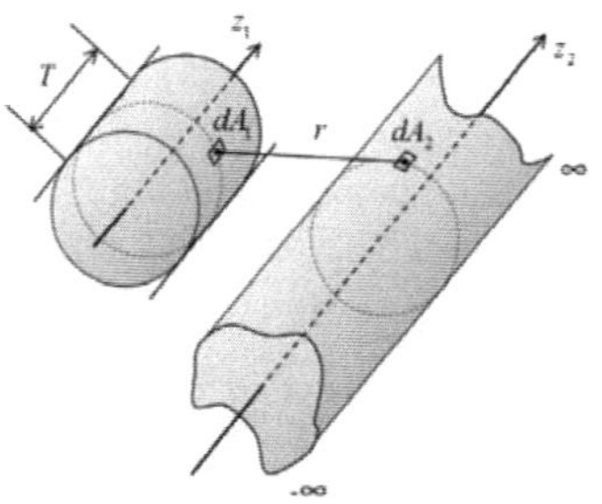

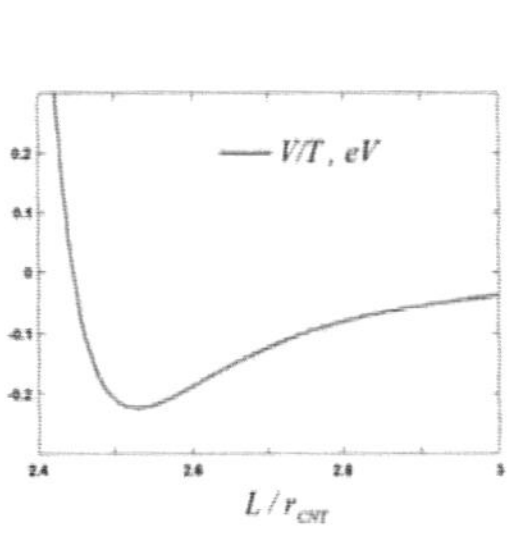

Generalized LJ potential for interaction between two segments:

$$V(R) = 4\varepsilon'' \left(\frac{A}{D^\alpha} - \frac{B}{D^\beta} \right) \quad D = R / r_{CNT} - 2$$

ITASCA Anderson, T. et al., 2010, Journal of Nanotech. in Eng. and Med. 1.

Van der Waals Pair Potentials

Use of the pair potentials leads to an artificial structure of intertube potential:

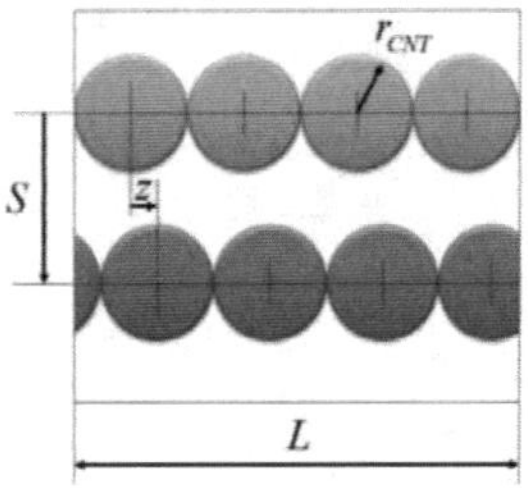

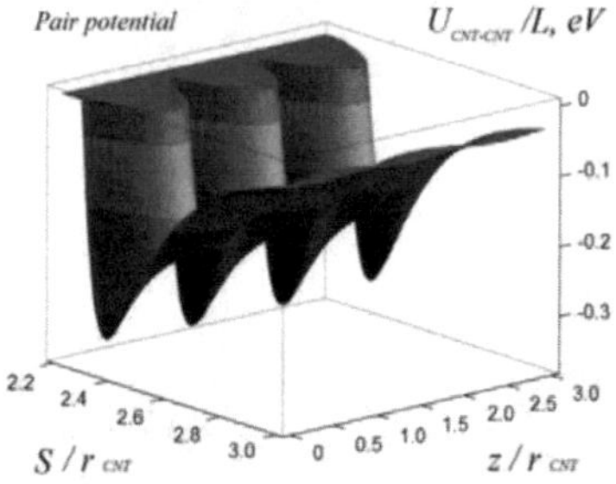

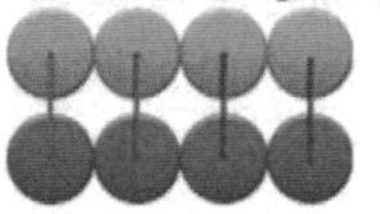

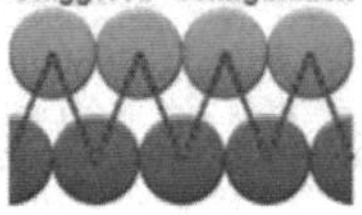

The artificial structure of the potential leads to a high adhesive shear strength between CNTs, which significantly affects mechanical response of CNT-based material on a macroscale, and the mobility of CNT in self-assembly simulations.

Van der Waals Smooth Translation

Introduce the anisotropy of the potential as a Fourier series with respect to angle θ with coefficients fitted to provide smooth intertube potential.

$$U(D,\theta) = U(D^*(\theta)) \quad D^*(\theta) = R/[r_{CNT}\Theta^k(\theta)] - 2 \quad \Theta^k(\theta) = 1 + \sum_{i=1}^{k} C_i((-1)^{i-1} + \cos(2i\theta))$$

k constants C_i are found by nonlinear fitting

k = 5:
C_1 = 0.34250
C_2 = 0.02720
C_3 = 0.00148
C_4 = -0.00014
C_5 = 0.00029

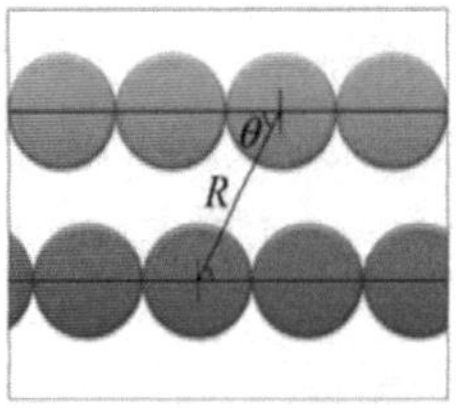

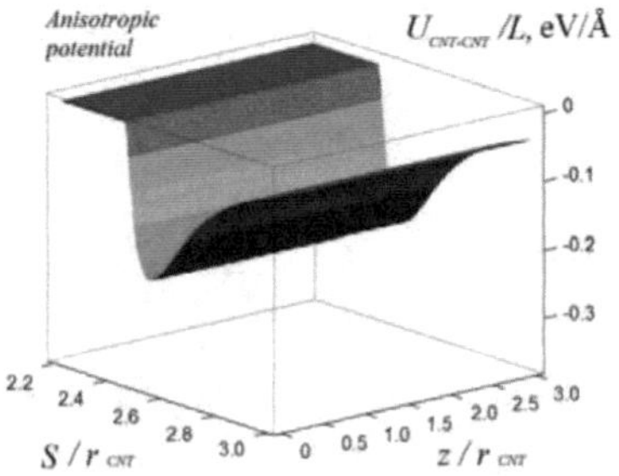

The resulting potential provides the smooth sliding of CNTs, high mobility of CNTs in self-assembly experiments.

Ostanin, I. Ballarini, R., Potyondy, D. and Dumitrica, R., 2012, JMPS, 61(3).

Outline

- Brief Introduction to the Discrete Element Method (DEM)
- DEM Modelling of Carbon Nanotubes (CNTs)
- CNT Characterization

- DEM Modelling of Magnetic Dipoles
- Self-Assembly Scoping Study
- Induced Magnetization Scoping Study

Applications of the developed model

- Analysis of the self-folded configurations of CNTs - CNT rings and CNT rackets
- Analysis of 1D CNT structures - CNT bundles, CNT ropes.
- Analysis 2D CNT structures - CNT papers, CNT films.

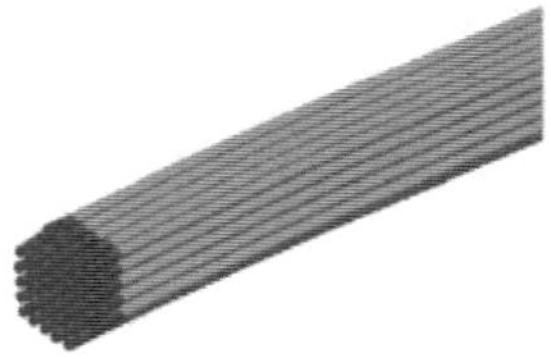

CNT bundle

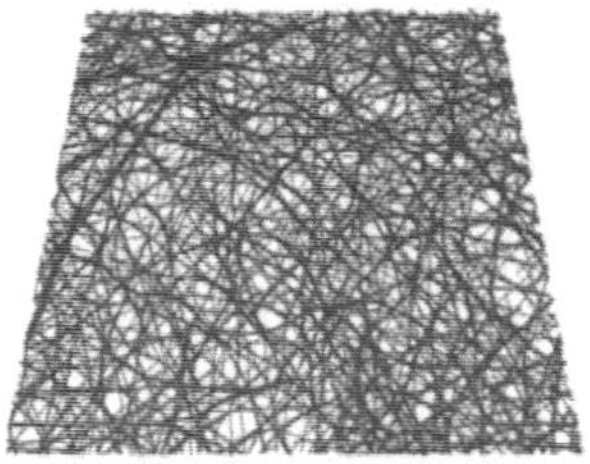

CNT paper

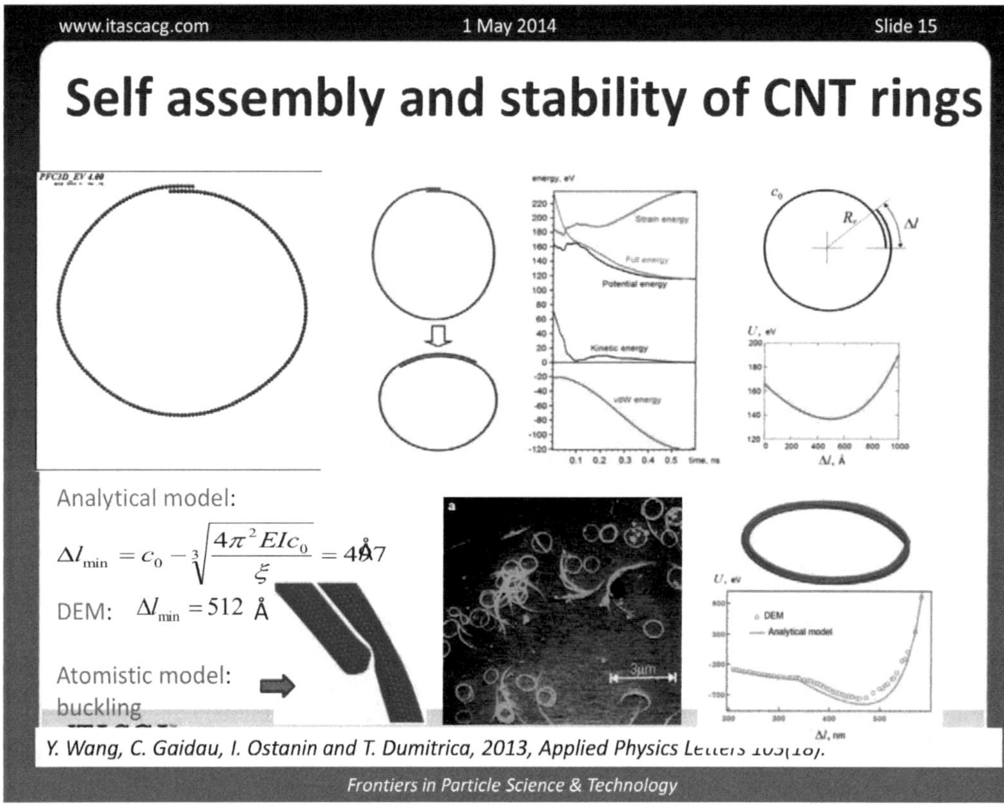
www.itascacg.com
1 May 2014
Slide 15
Self assembly and stability of CNT rings
Strain energy
Full energy
Potential energy
Kinetic energy
vdW energy
time, ns
c_0
R_0
Δl
Δl, Å
Analytical model:
$\Delta l_{\min} = c_0 - \sqrt[3]{\frac{4\pi^2 EIc_0}{\xi}} = 4Å7$
DEM: $\Delta l_{\min} = 512$ Å
Atomistic model:
buckling
3μm
DEM
Analytical model
Δl, nm
Y. Wang, C. Gaidau, I. Ostanin and T. Dumitrica, 2013, Applied Physics Letters 103(18).
Frontiers in Particle Science & Technology

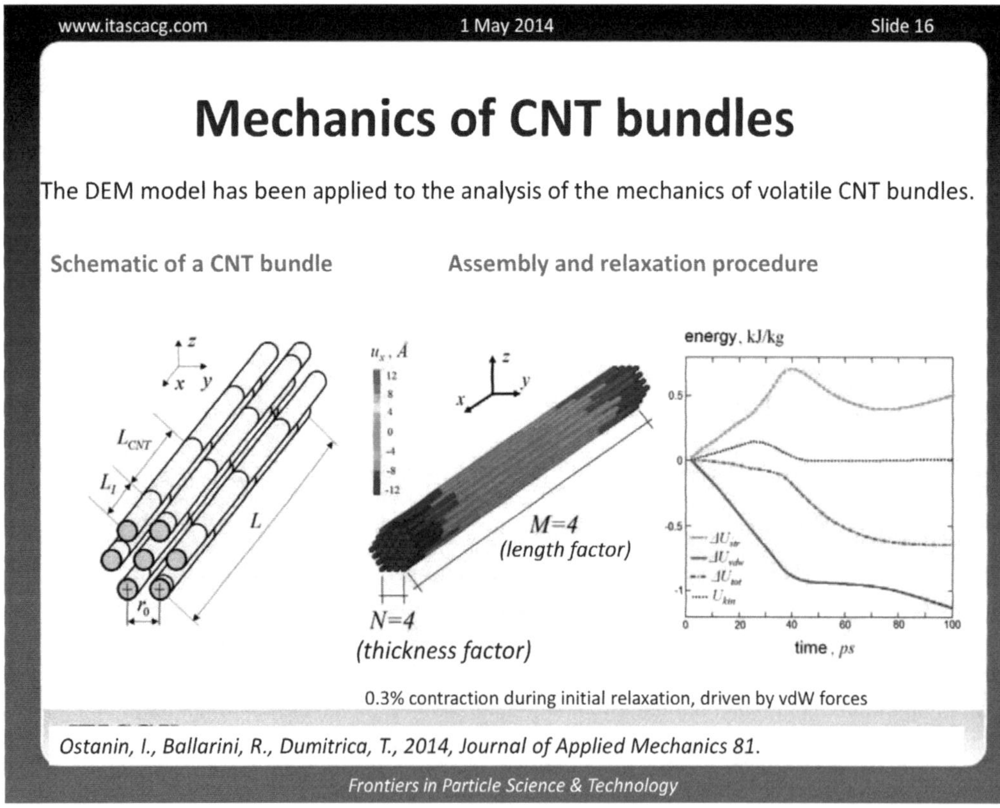
www.itascacg.com
1 May 2014
Slide 16
Mechanics of CNT bundles
The DEM model has been applied to the analysis of the mechanics of volatile CNT bundles.
Schematic of a CNT bundle
Assembly and relaxation procedure
L_{CNT}
L_l
L
r_0
u_x, Å
M=4
(length factor)
N=4
(thickness factor)
energy, kJ/kg
ΔU_{str}
ΔU_{vdw}
ΔU_{tot}
U_{kin}
time, ps
0.3% contraction during initial relaxation, driven by vdW forces
Ostanin, I., Ballarini, R., Dumitrica, T., 2014, Journal of Applied Mechanics 81.
Frontiers in Particle Science & Technology

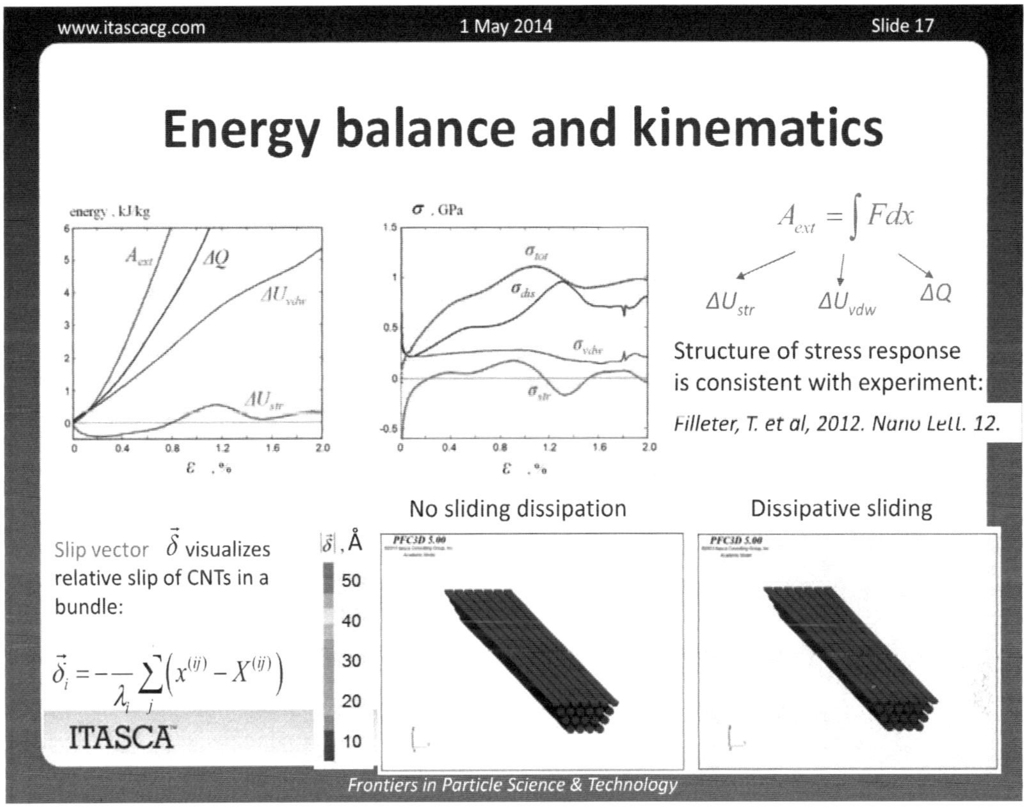
www.itascacg.com
1 May 2014
Slide 17
Energy balance and kinematics
energy, kJ/kg
A_{ext}
ΔQ
ΔU_{vdw}
ΔU_{str}
ε, %
σ, GPa
σ_{tot}
σ_{dis}
σ_{vdw}
σ_{str}
$A_{ext} = \int F dx$
ΔU_{str}
ΔU_{vdw}
ΔQ
Structure of stress response is consistent with experiment:
Filleter, T. et al, 2012. Nano Lett. 12.
No sliding dissipation
Dissipative sliding
Slip vector $\vec{\delta}$ visualizes relative slip of CNTs in a bundle:
$\vec{\delta}_i = -\frac{}{\lambda_i} \sum_j \left(x^{(ij)} - X^{(ij)} \right)$
$|\vec{\delta}|$, Å
50
40
30
20
10
PFC3D 5.00
ITASCA
Frontiers in Particle Science & Technology

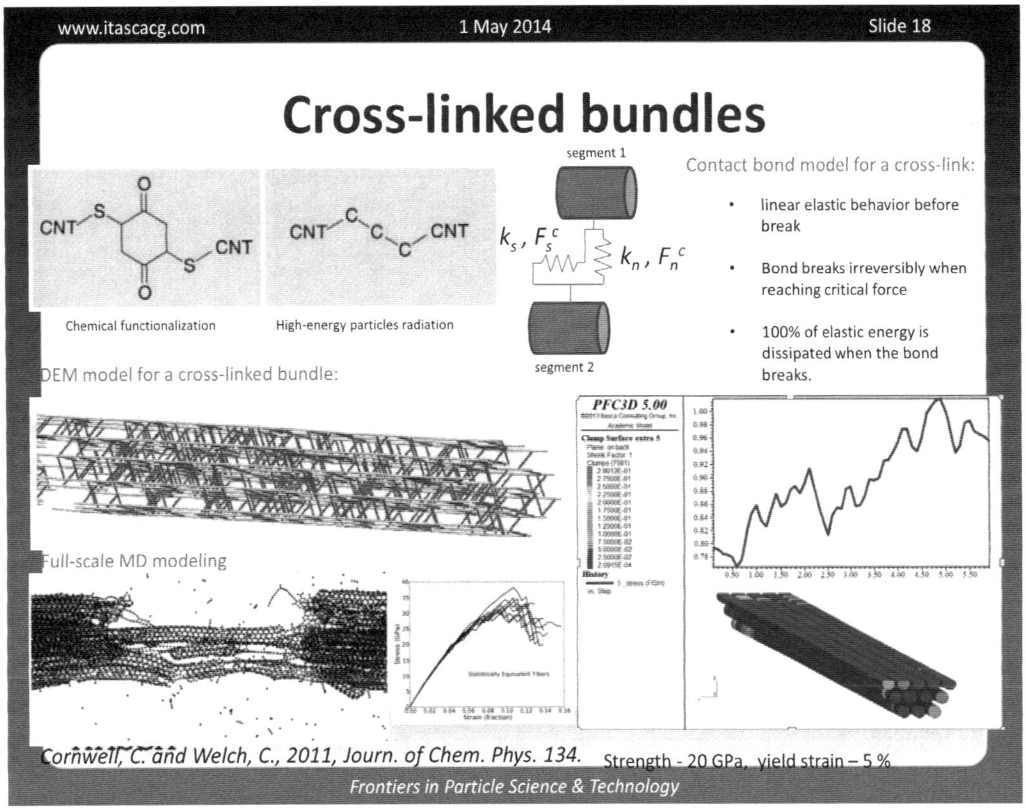
www.itascacg.com
1 May 2014
Slide 18
Cross-linked bundles
segment 1
segment 2
k_s, F_s^c
k_n, F_n^c
CNT
S
O
C
Chemical functionalization
High-energy particles radiation
Contact bond model for a cross-link:
• linear elastic behavior before break
• Bond breaks irreversibly when reaching critical force
• 100% of elastic energy is dissipated when the bond breaks.
DEM model for a cross-linked bundle:
Full-scale MD modeling
PFC3D 5.00
Cornwell, C. and Welch, C., 2011, Journ. of Chem. Phys. 134.
Strength - 20 GPa, yield strain – 5 %
Frontiers in Particle Science & Technology

Mechanical test on a CNT film

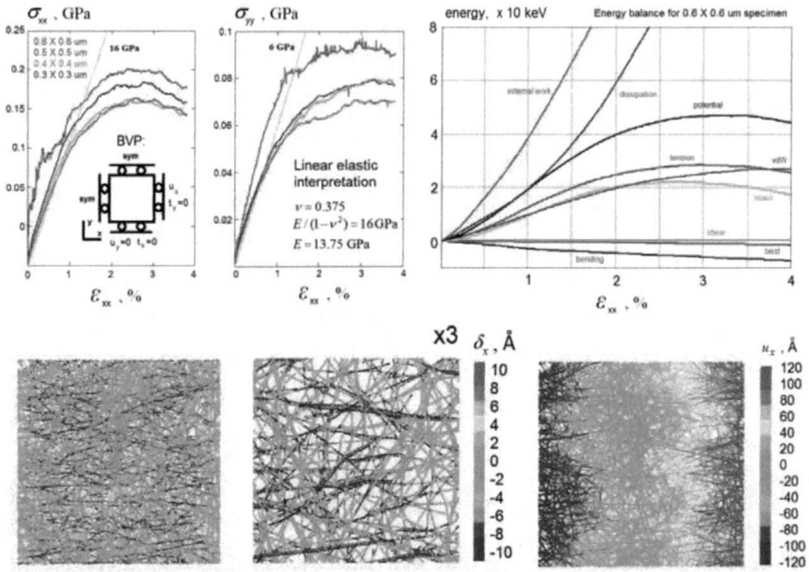

Outline

- Brief Introduction to the Discrete Element Method (DEM)
- DEM Modelling of Carbon Nanotubes (CNTs)
- CNT Characterization
- DEM Modelling of Magnetic Dipoles
- Self-Assembly Scoping Study
- Induced Magnetization Scoping Study

ITASCA

www.itascacg.com 1 May 2014 Slide 21

Motivation

ITASCA™

Frontiers in Particle Science & Technology

www.itascacg.com 1 May 2014 Slide 22

Dipole Approximation - Convergence to Analytic Solution

$$\mathbf{F}_k = \frac{\mu_0}{4\pi} \sum_{i=1}^{N} \frac{1}{|\mathbf{r}_k - \mathbf{r}_i|^4} \Big[-15\mathbf{n}_{ik}\big((\mathbf{m}_k \cdot \mathbf{n}_{ik})(\mathbf{m}_i \cdot \mathbf{n}_{ik})\big) + 3\mathbf{n}_{ik}(\mathbf{m}_k \cdot \mathbf{m}_i) + 3\big(\mathbf{m}_k(\mathbf{m}_i \cdot \mathbf{n}_{ik}) + \mathbf{m}_i(\mathbf{m}_k \cdot \mathbf{n}_{ik})\big)\Big] \quad (8)$$

with $\mathbf{n}_{ik} = (\mathbf{r}_k - \mathbf{r}_i)/|\mathbf{r}_k - \mathbf{r}_i|$, and accordingly the torque $\mathbf{T}_k$,

$$\mathbf{T}_k = \frac{\mu_0}{4\pi} \sum_{i=1}^{N} \left[\frac{3(\mathbf{m}_k \times \mathbf{n}_{ik})(\mathbf{m}_i \cdot \mathbf{n}_{ik}) - \mathbf{m}_k \times \mathbf{m}_i}{|\mathbf{r}_k - \mathbf{r}_i|^3} \right]. \quad (9)$$

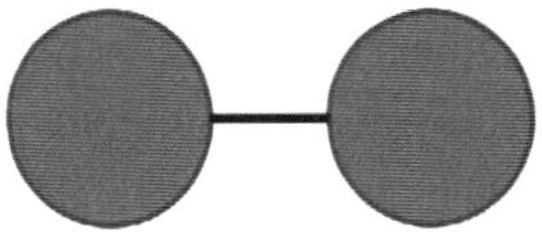

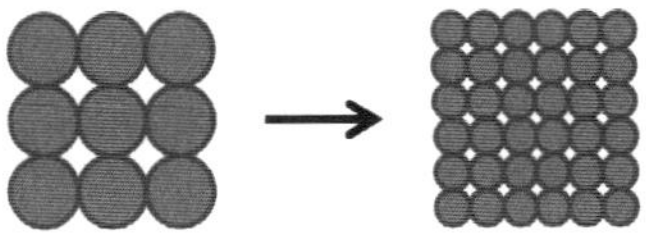

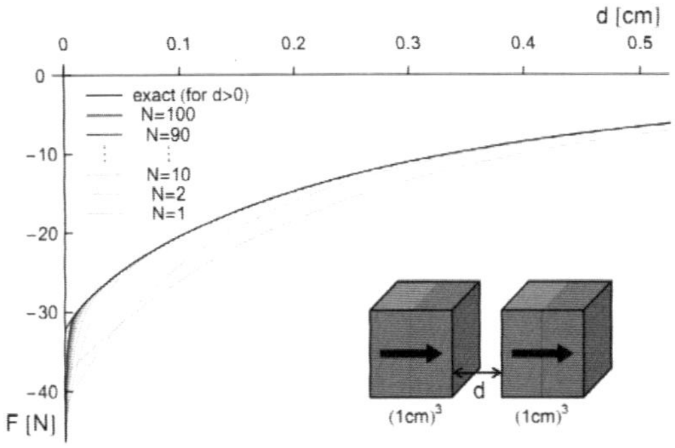

ITASCA™

Frontiers in Particle Science & Technology

www.itascacg.com 1 May 2014 Slide 23

Simple Verification

ITASCA

Frontiers in Particle Science & Technology

www.itascacg.com 1 May 2014 Slide 24

Outline

- Brief Introduction to the Discrete Element Method (DEM)
- DEM Modelling of Carbon Nanotubes (CNTs)
- CNT Characterization
- DEM Modelling of Magnetic Dipoles
- Self-Assembly Scoping Study

- Induced Magnetization Scoping Study

ITASCA

Frontiers in Particle Science & Technology

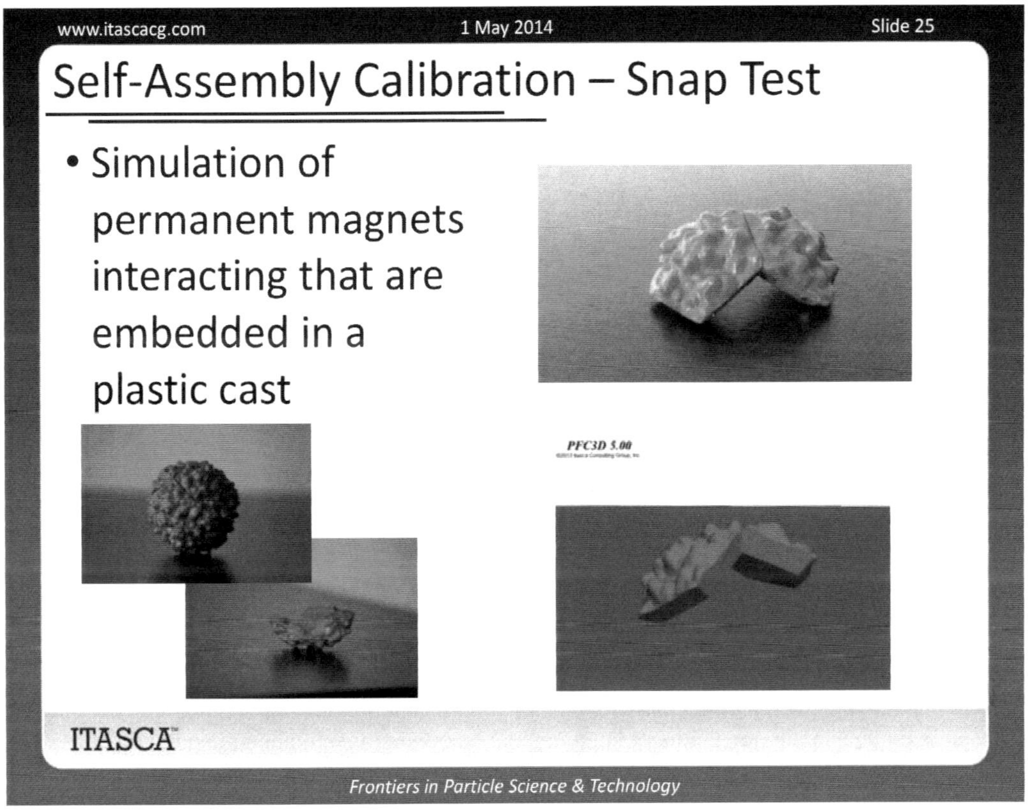
www.itascacg.com
1 May 2014
Slide 25
Self-Assembly Calibration – Snap Test
• Simulation of permanent magnets interacting that are embedded in a plastic cast
PFC3D 5.00
ITASCA
Frontiers in Particle Science & Technology

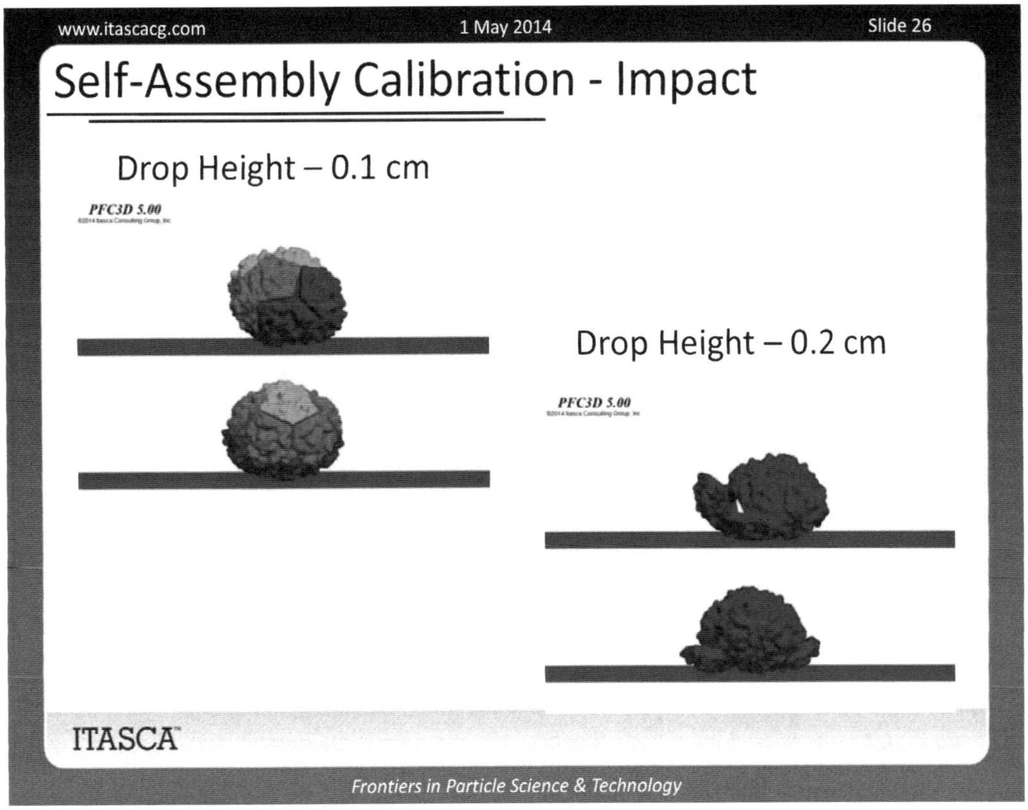
www.itascacg.com
1 May 2014
Slide 26
Self-Assembly Calibration - Impact
Drop Height – 0.1 cm
PFC3D 5.00
Drop Height – 0.2 cm
PFC3D 5.00
ITASCA
Frontiers in Particle Science & Technology

Self-Assembly Simulation

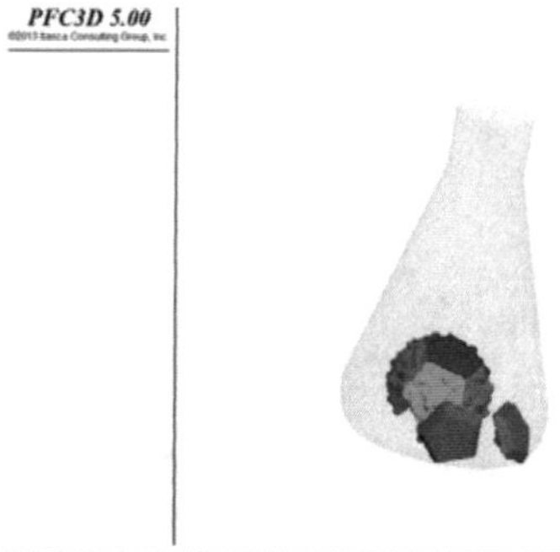

ITASCA™

Outline

- Brief Introduction to the Discrete Element Method (DEM)
- DEM Modelling of Carbon Nanotubes (CNTs)
- CNT Characterization
- DEM Modelling of Magnetic Dipoles
- Self-Assembly Scoping Study
- Induced Magnetization Scoping Study

ITASCA™

Simple Model and Calibration

$$\mathbf{M}_i = \frac{3}{\mu_0}\frac{\chi}{1+\chi}\,\mathbf{B}(\mathbf{r}_i)\,.$$

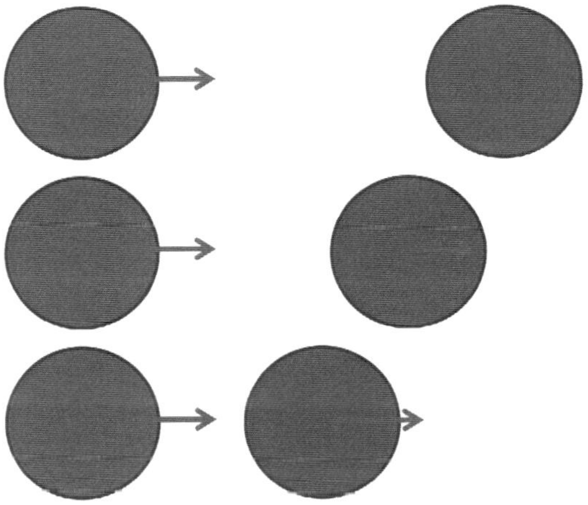

ITASCA™

Magnetic Drum Separator

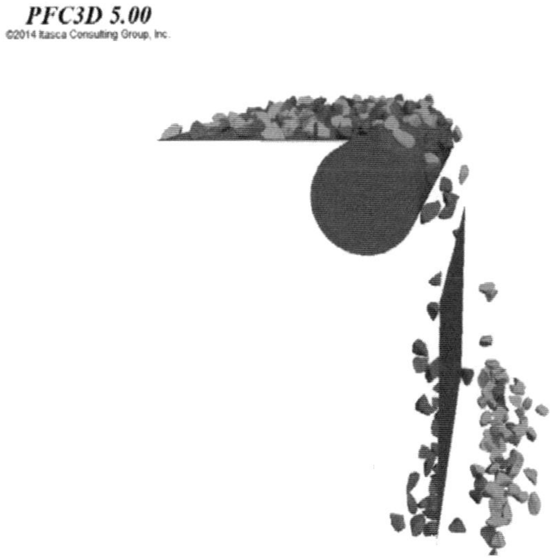

ITASCA™

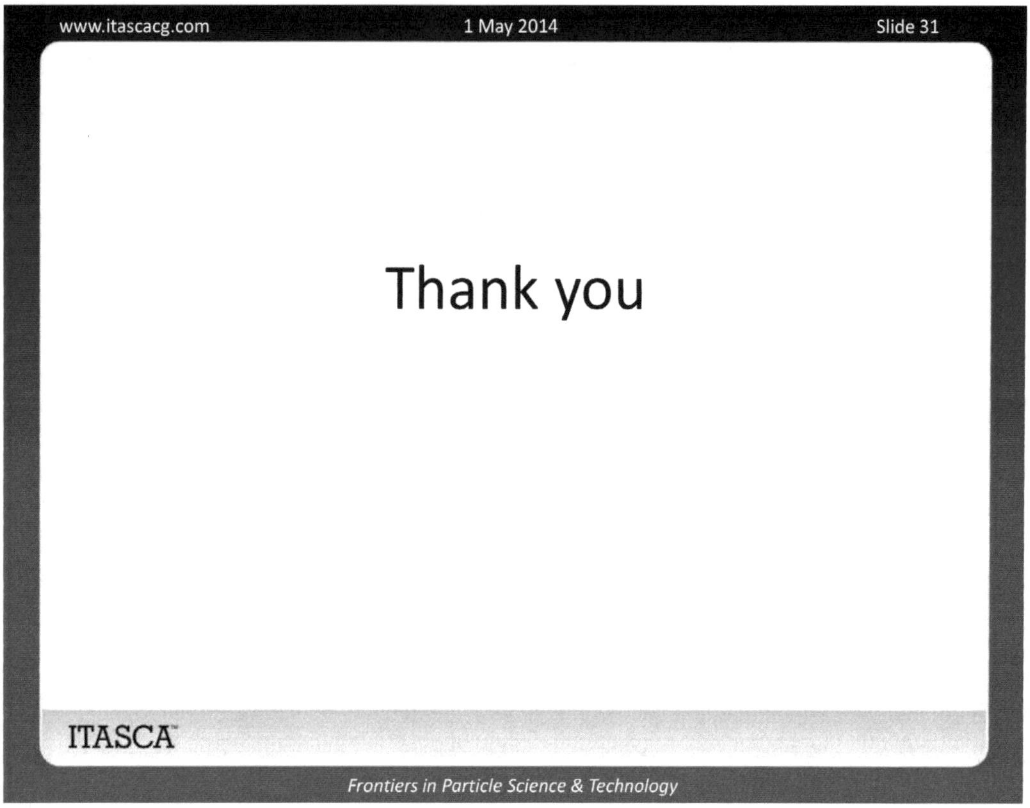

www.itascacg.com
1 May 2014
Slide 31
Thank you
ITASCA
Frontiers in Particle Science & Technology

Granular Jamming and Flow

R.P. Behringer

Frontiers in Particle Science and Technologhy

April 30, 2014

Support: NSF, NASA, BSF, ARO, DTRA, IFPRI

Collaborators: Max Bi, Nicolas Brodu, Joshua Dijksman, Dan Howell, Trush Majmudar, Jie Ren, Sumantra Sarkar, Junyao Tang, Jie Zhang, Hu Zheng, Bulbul Chakraborty , Lou Kondic, Miro Kramer, Konstantin Mischaikov, David Schaeffer, Antoinette Tordesillas

Abe Clark, Corentin Coulais, Julien Dervaux, Tyler Earnest, Somayeh Farhadi, Junfei Geng, Guillaume Reydellet, Sarath Tennakoon, Brian Tighe, John Wambaugh, Brian Utter, Peidong Yu, Eric Clément, Karin Dahmen, Karen Daniels, Olivier Dauchot, Isaac Goldhirsch, Wolfgang Losert, Jackie Krim, Stefan Luding, Chris Marone, Guy Metcalfe, Paul Mort, Corey O'Hern, Josh Socolar, Matthias Sperl

Roadmap

Use experiments to explore:

- Key aspects of dense granular materials
- Jamming dynamics of frictional systems under compression and shear
- Dynamics near shear jamming
- Impacts

Context for Dense Granular Phases—some identifying features

Frozen structure: forces are carried preferentially on force chains, 'long' force chains → anisotropy in local stress

Deformation, particularly shear, leads to large spatio-temporal fluctuations

Also, shear leads to dilation and shear bands

Granular materials jam, and shear leads to jamming below φ_J,

Important for dense flows

Friction and preparation history matter

GM's exhibit novel meso-scopic structures:

Force Chains:

These depend on the preparation of the material—long chains → shear → stress and network anisotropy

2d Shear → Experiment

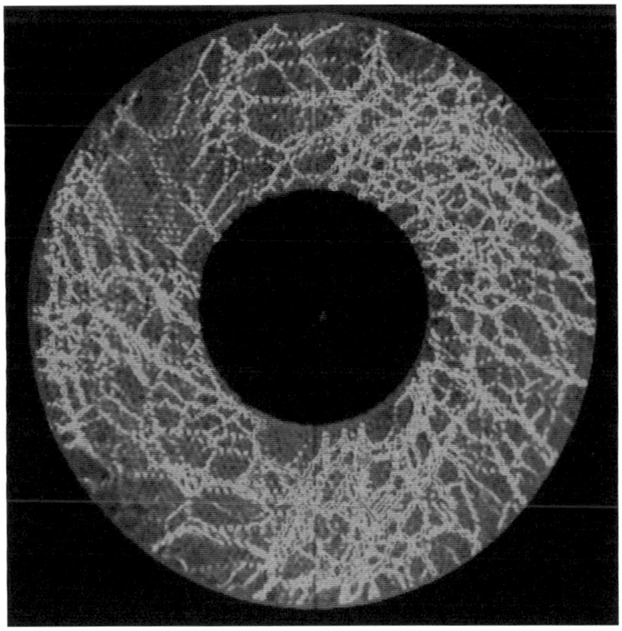

Howell et al.
PRL 82, 5241 (1999)

Anisotropy in force chains can lead to surprising results—example sand piles

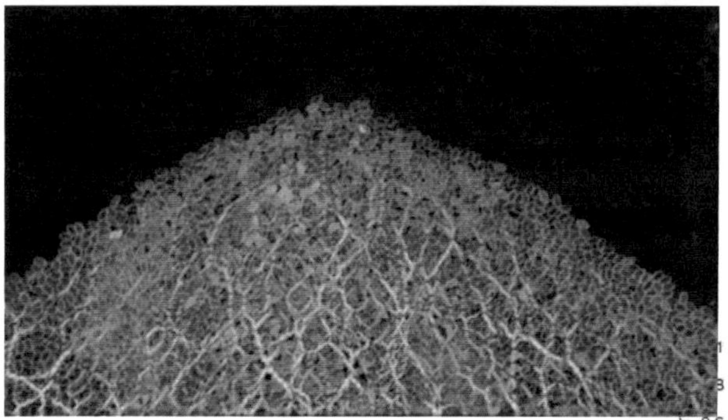

←2D photoelastic image (Geng et al.)

3D heaps and pressure→ at base (Vanel et al.)

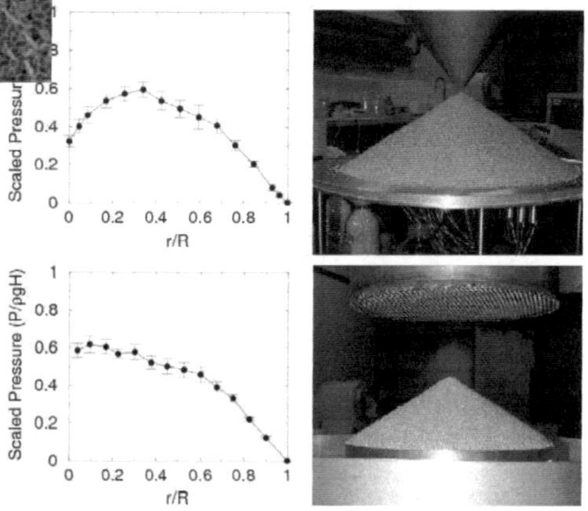

One frame, showing jam and force chain arch

2D hopper flow

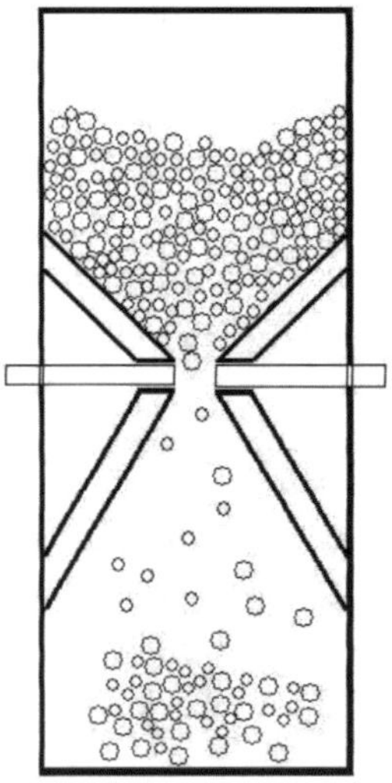

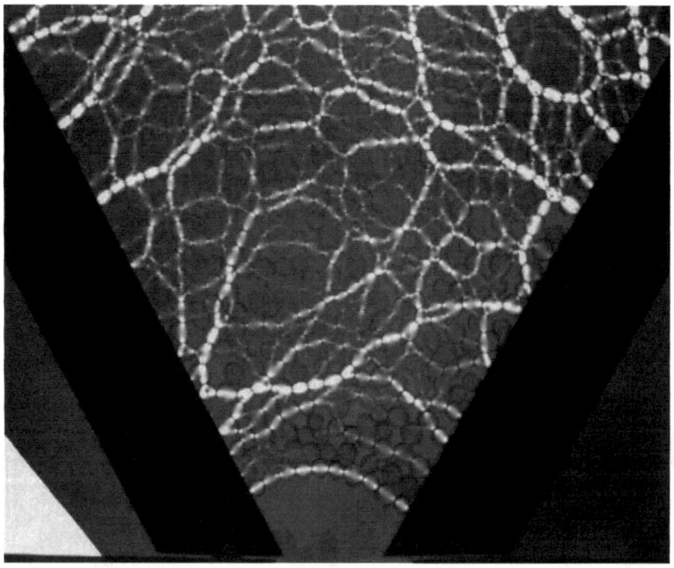

In hopper flows—formation of force chains arches at outlet leads to jamming (Junyao Tang)

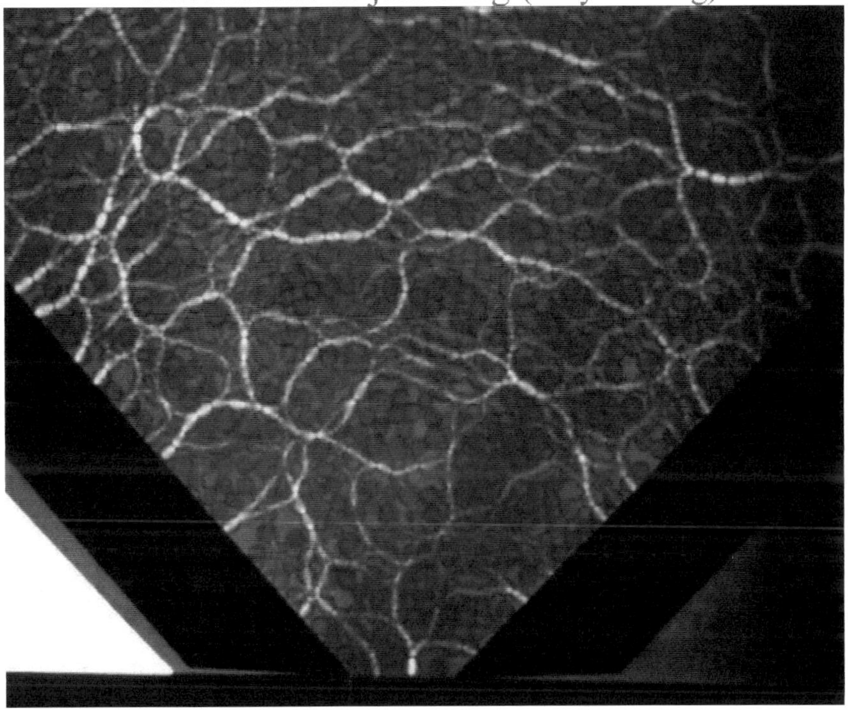

4) Granular materials jam

- Jamming—how disordered N-body systems becomes solid-like as particles are brought into contact, or fluid-like when grains are separated—thought to apply to many systems, including GM's foams, colloids, glasses…
- Isotropic case: Density is implicated as a key parameter, expressed as packing (solid fraction) φ
- Marginal stability (isostaticity) for spherical particles (disks in 2D) contact number, Z, attains a critical value, Z_{iso} at φ_{iso}
- Z_{iso} depends on dimension, friction.

But, experiments on frictional particles show other interesting behavior

Two kinds of state, depending on φ

1) ...$\varphi_S < \varphi < \varphi_J$—states arise under shear, $|\tau| > 0$
2) ...$\varphi > \varphi_J$—jammed states occur at $\tau = 0$

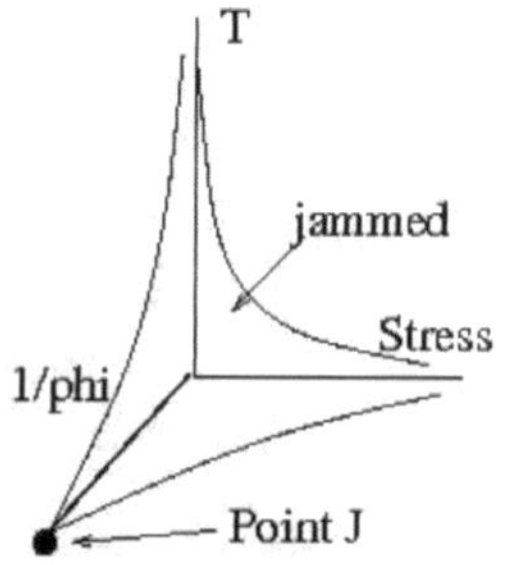

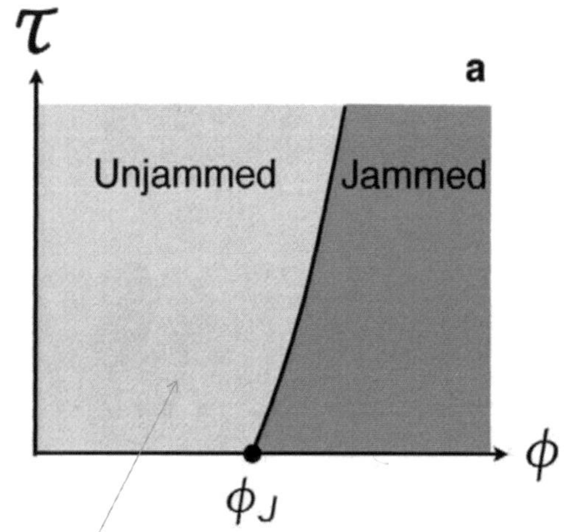

Original (Liu &Nagel, Nature 1998)

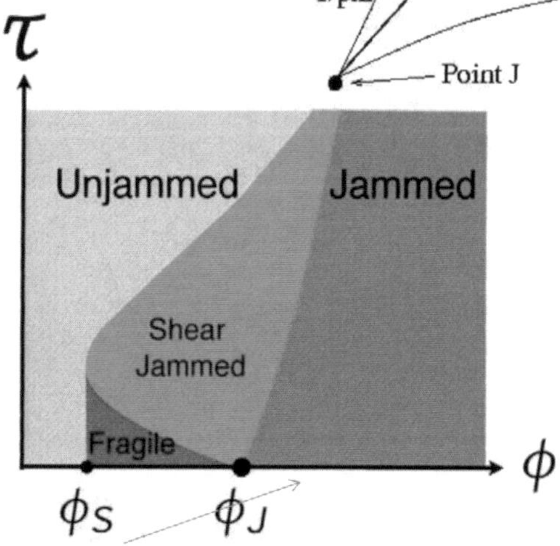

Bi et al. Nature, 480, 355 (2011)

Key new approach: obtain grain contact forces

Experiment--raw

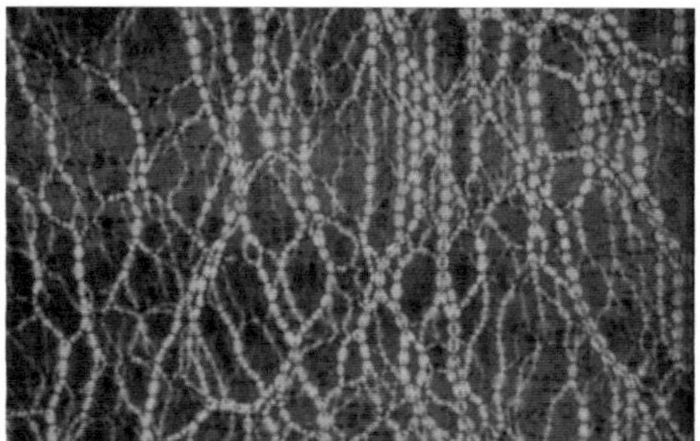

Experiment Color filtered

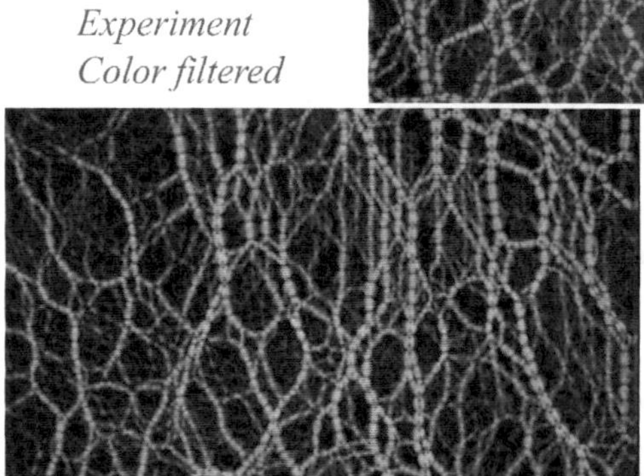

Reconstruction From force inverse algorithm

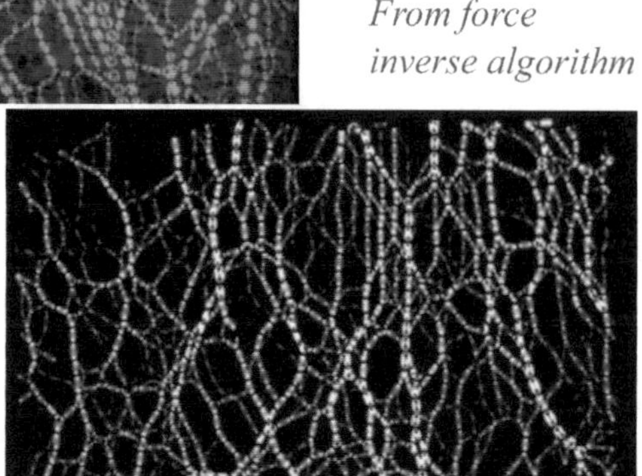

Obtaining stresses and fabric from experimental data

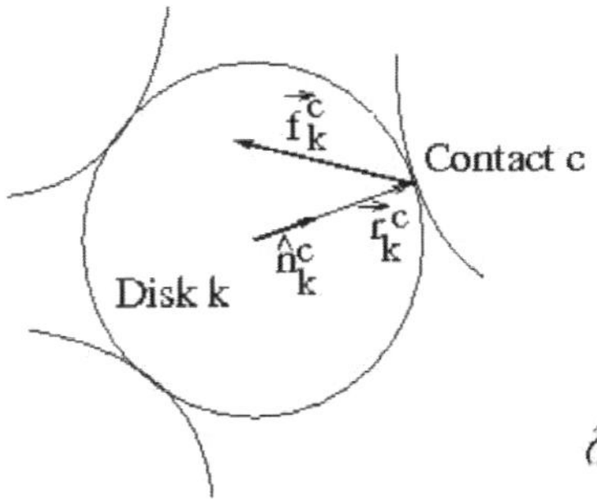

$$\hat{\sigma} = \frac{1}{V}\sum_{i\neq j}\vec{r}_{ij}\otimes\vec{f}_{ij}, \qquad \text{Stress}$$

$$\hat{R} = \frac{1}{N}\sum_{i\neq j}\frac{\vec{r}_{ij}}{\|\vec{r}_{ij}\|}\otimes\frac{\vec{r}_{ij}}{\|\vec{r}_{ij}\|}, \qquad \text{Fabric}$$

These quantities can be coarse-grained to produce continuum fields

Experiment 1: Biaxial Tester

Control spacing between opposing pairs of walls

Experiments use biaxial tester→ and photoelastic particles

Isotropic compression

Experiments on frictional grains resemble predictions for frictionless grains

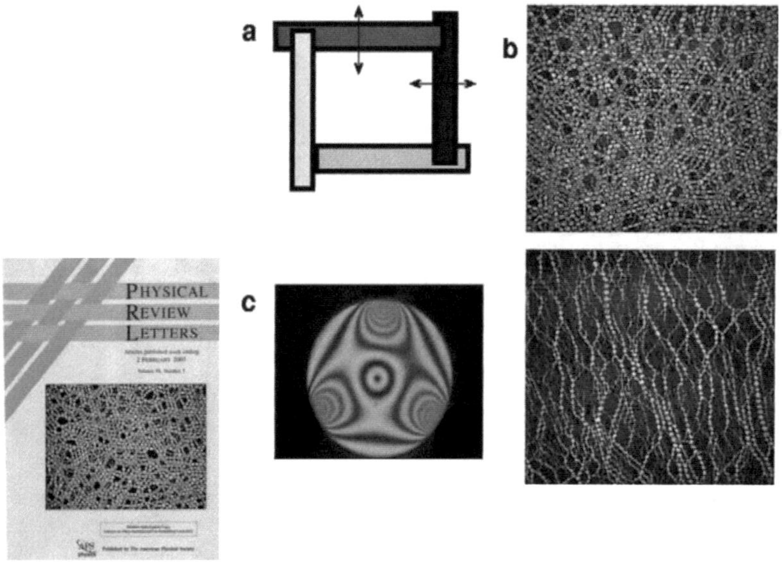

← *Isotropic compression*

← *Pure shear*

Majmudar, Sperl, Luding and BB, PRL 2007

Three predictions for the Isotropic Jamming Transition

1) $Z \sim Z_I + (\varphi - \varphi_c)^{\acute{\alpha}}$

(discontinuity)

Exponent $\acute{\alpha} \approx 1/2$

2) $\boldsymbol{P \sim (\varphi - \varphi_c)^{\beta}}$

β depends on force law (= 1 for ideal disks)

3) Henkes and Chakraborty: entropy-based model: P(Z)

Z = contacts/particle; Φ = packing fraction

Predictions (e.g. O'Hern, Silbert, Liu and Nagel; Torquato et al., Schwarz et al., Henkes and Chakraborty

LSQ Fits for Z give an exponent of 0.5 to 0.6

$\varphi_c \approx 0.84$

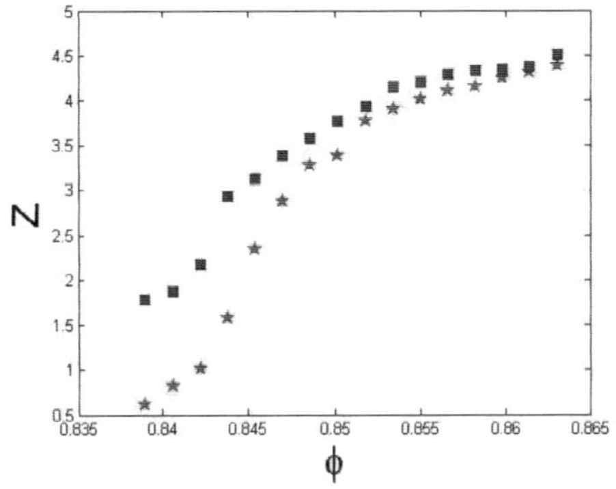

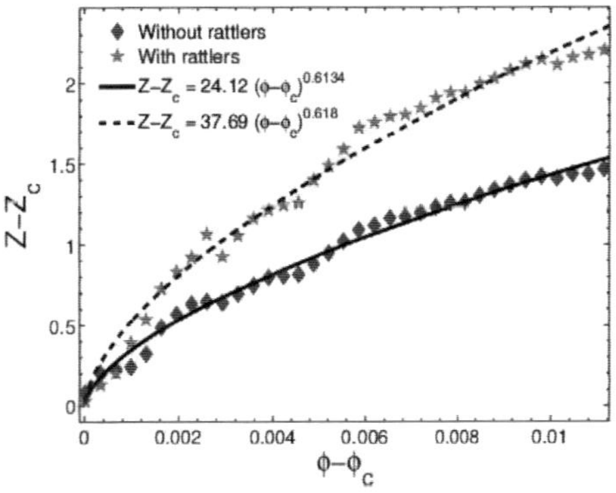

Transition is 'rapid' but not discontinuous—return to this

LSQ Fits for P give $\beta \approx 1.0$ to 1.1

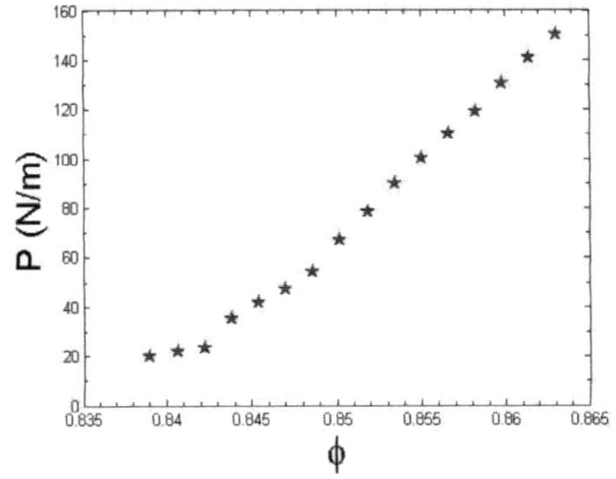

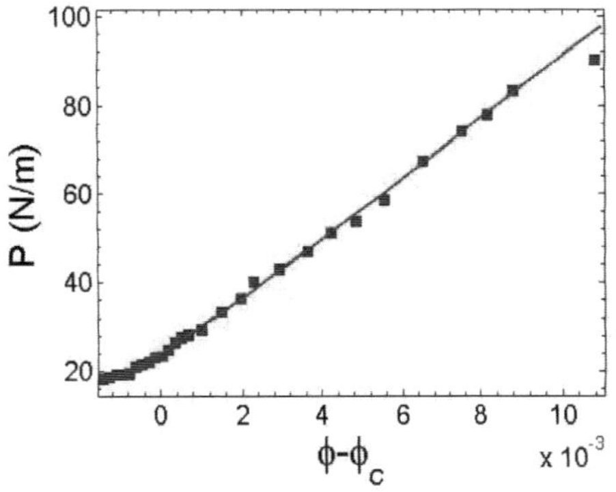

Jamming under Shear (role of friction)

How do disordered solids lose/gain their solidity under shear?

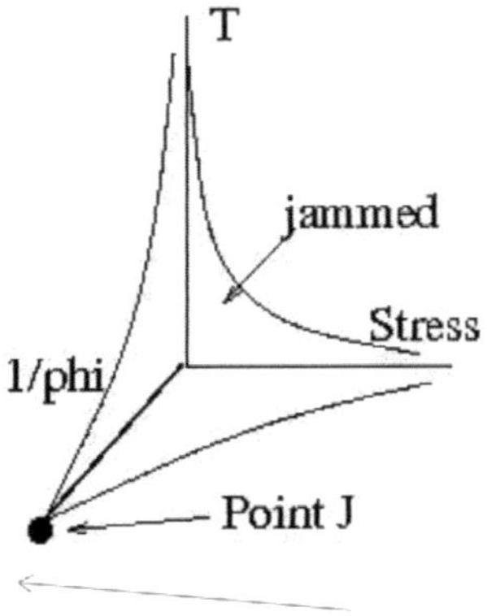

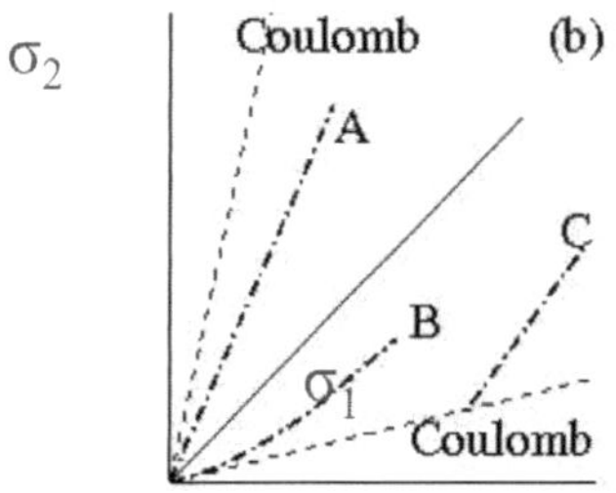

What happens here, when shear is applied to a granular material?

Note: $P = (\sigma_2 + \sigma_1)/2$ Shear stress: $\tau = (\sigma_2 - \sigma_1)/2$

Coulomb failure: $|\tau|/P = \mu$

Return to Biaxial Experiments: Choose pure shear mode

Biax schematic

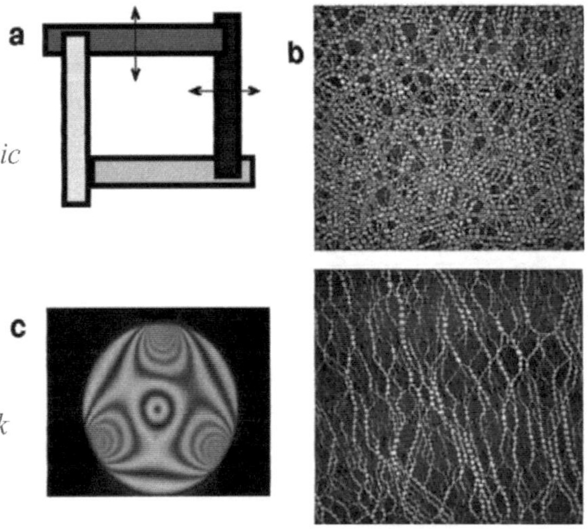

Compression

Image of Single disk

← Shear

~2500 particles, bi-disperse, d_L*=0.9cm,* d_S*= 0.8cm,* $N_S/N_L = 4$

Zhang, Majmudar, Tordesillas, BB, Granular Matter, 2010
Bi, Zhang, Chakraborty, BB, Nature, 2011

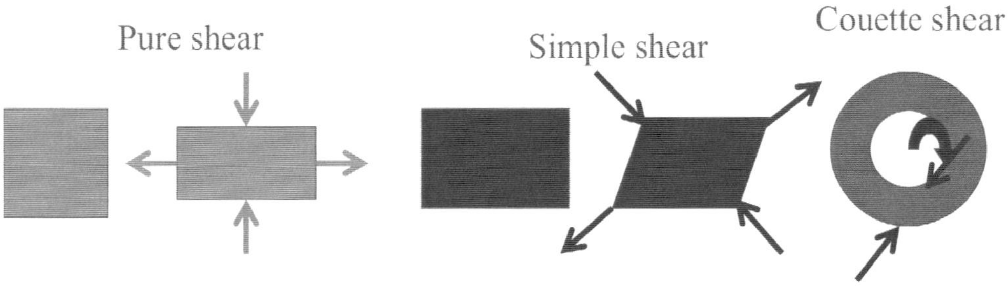

Different ways to apply shear:

Common feature of different protocols for shear

One compressive and one dilational direction
(no change of area)

First: Pure Shear Experiments using Biax

Time-lapse video (one pure shear cycle) shows force network evolution

Initial and final states following a shear cycle—no change in area

← Initial state, isotropic, no stress

States at the same density are differentiated by properties other than their density

Final state → large stresses

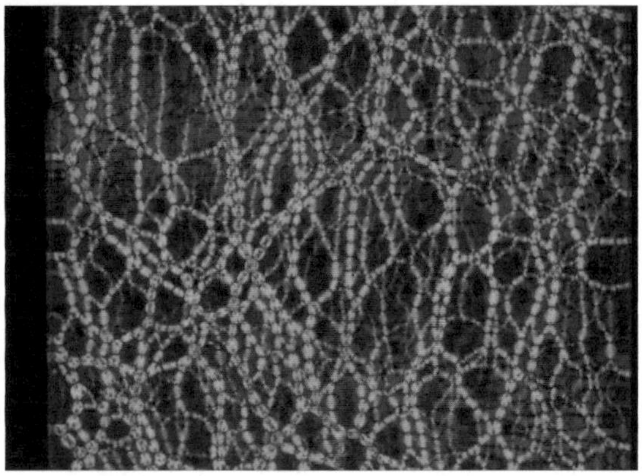

Pure shear

Simple shear

Couette shear

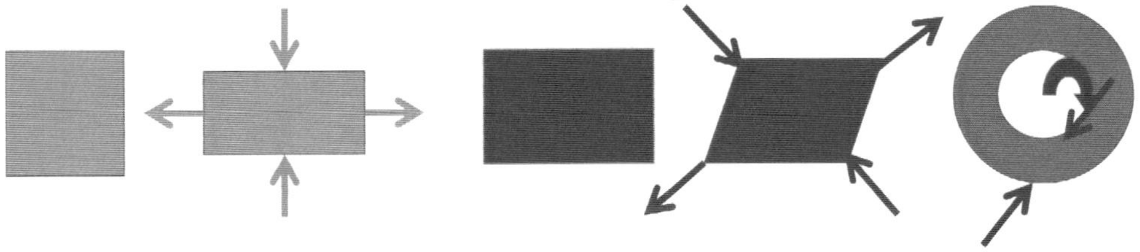

Different ways to apply shear:

Common feature of different protocols for shear

One compressive and one dilational direction
(no change of area)

2nd apparatus: quasi-uniform simple shear

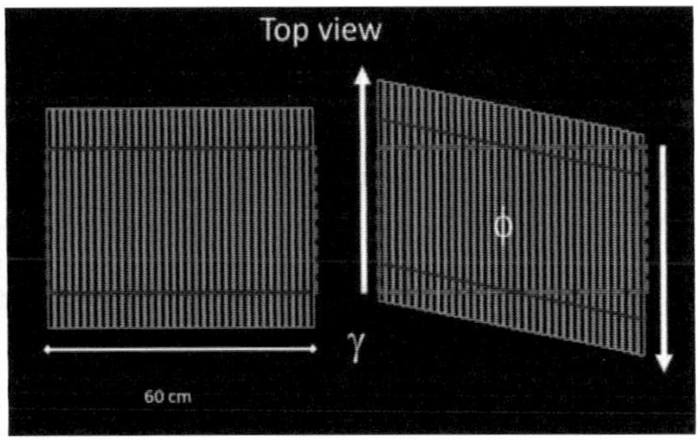

- Use bi-disperse particle:
  - d_S=0.5", d_L=0.65", N_S/N_L~3.3
  - Total particle number: ~1000
- Rectangle width ~ $25d_S$
- Slat width=0.5", comparable to particle size
- Shear strain $\varepsilon=x/x_0$, increases by 0.482% per step
- Take photos of both the normal view of the particles, and the polarized view of force chain structures.

(Jie Ren, Joshua Dijksman)

Ren et al.
PRL 110, 018302
(2013)

This new experimental approach supplies uniform shear

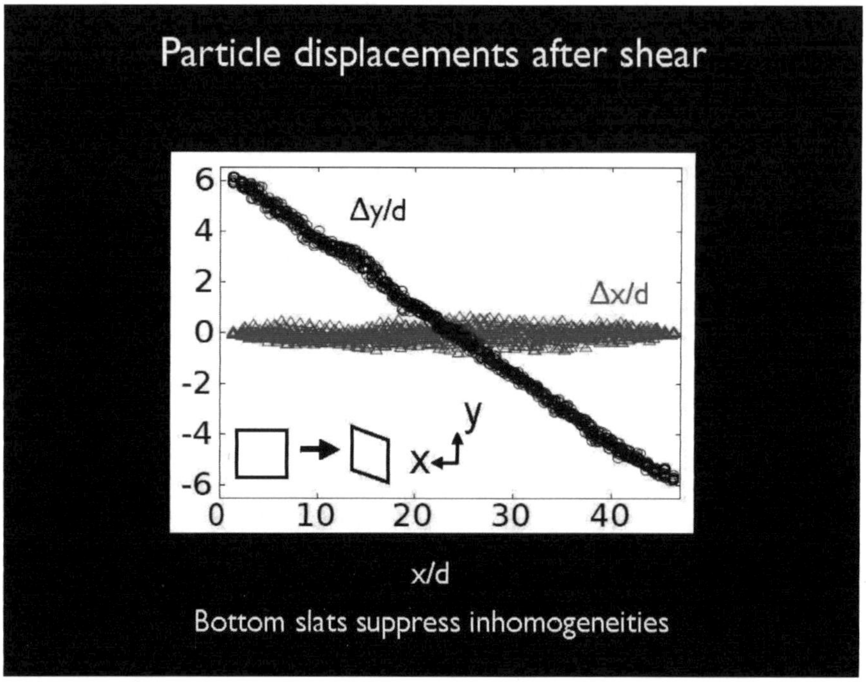

Uniform shear
No shear banding

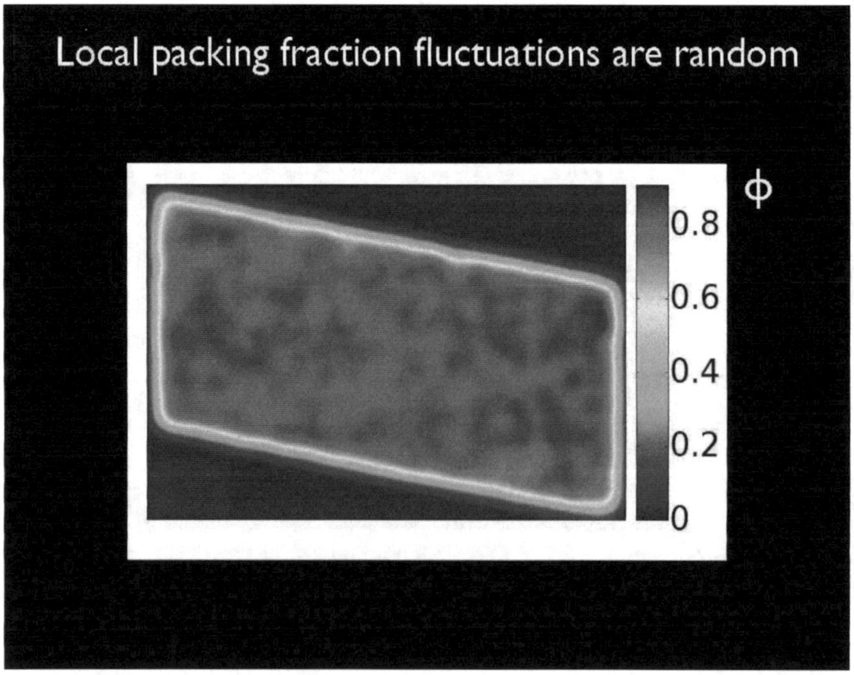

Time-Lapse Video of Shear-Jamming

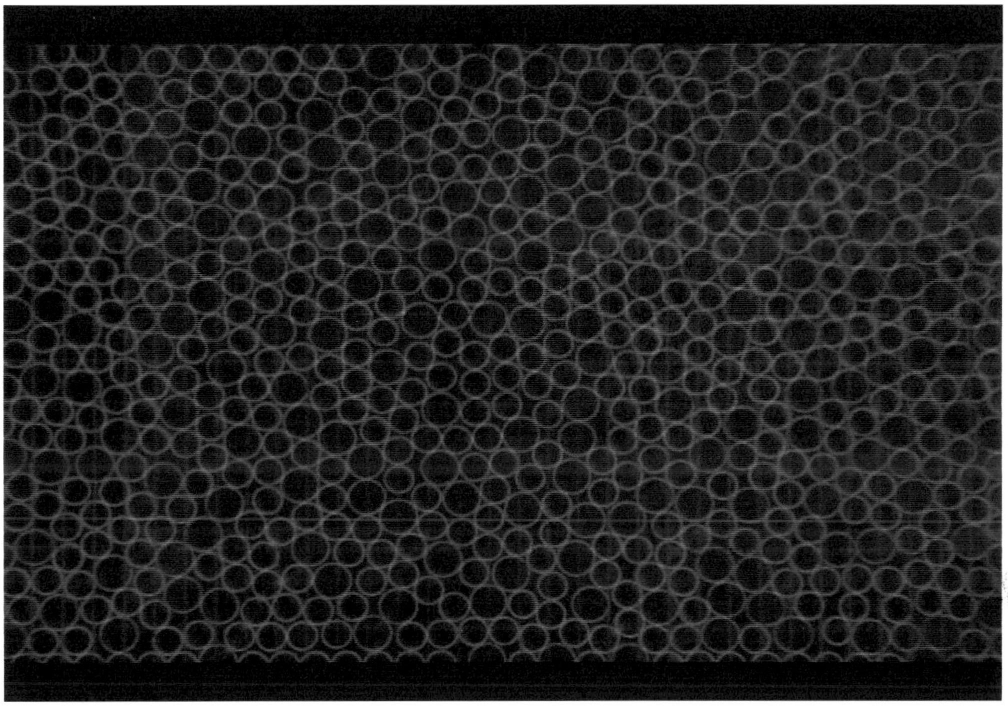

Range of densities for which shear jamming can be achieved

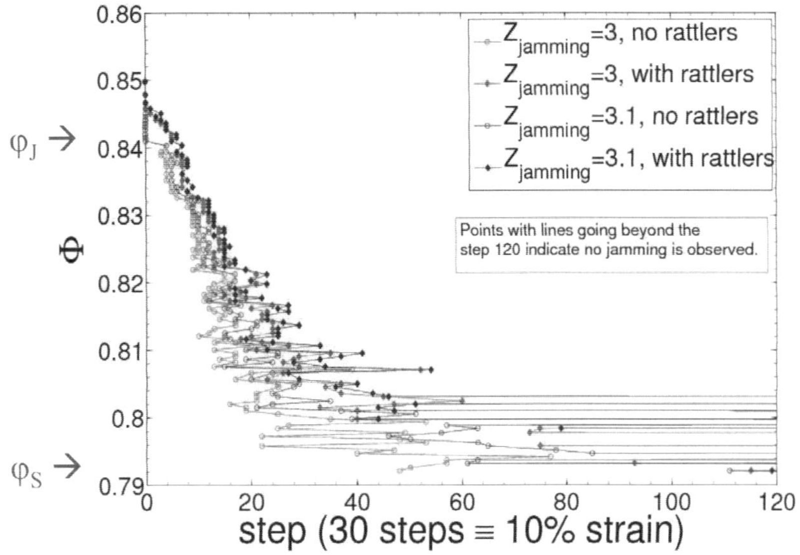

But: networks are key to shear jamming

Increasing shear strain—first unidirectional, then all-directional percolation of strong force network

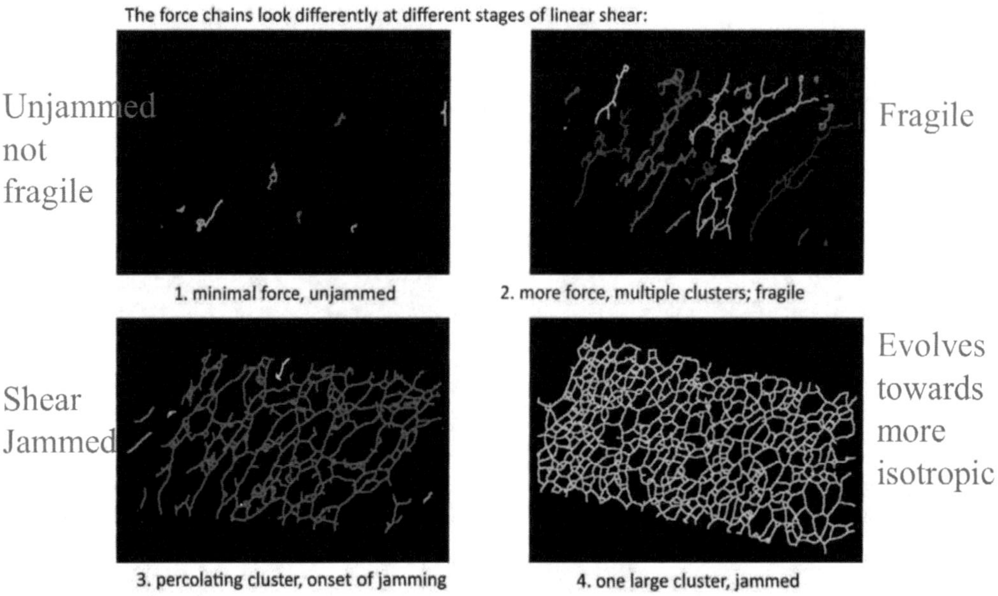

Jamming diagram for Frictional Particles

Three kinds of state, depending on φ and shear strain

1) …$\varphi_S < \varphi < \varphi_J$—for small shear, fragile states
2) …then with enough strain are jammed states
3) …$\varphi > \varphi_J$—jammed states occur at $\tau = 0$

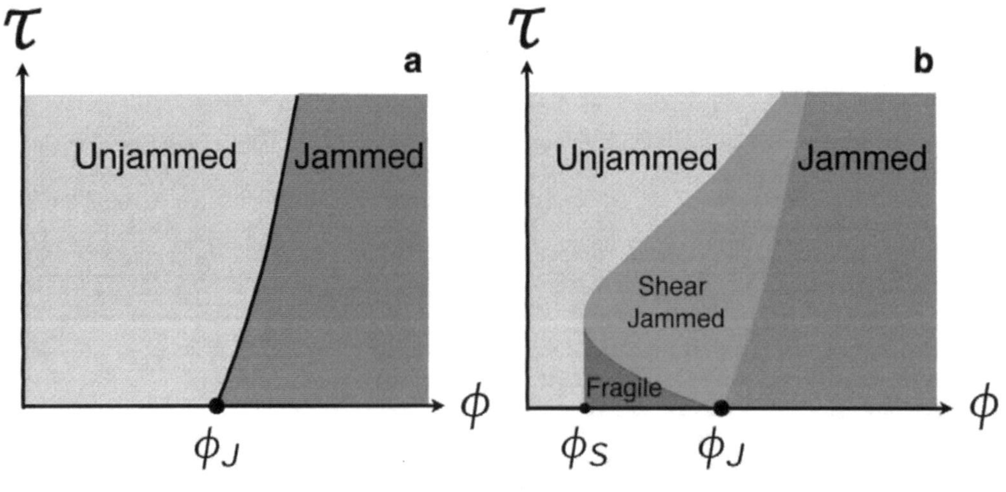

Original

New (Frictional)

Return to stress evolution with strain—generalized elasticity

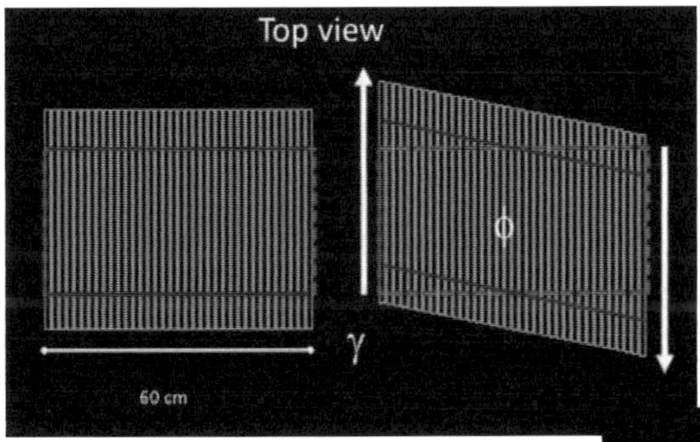

Augment unidirectional
Shear with cyclic shear—
Probes stress evolution
And dynamics

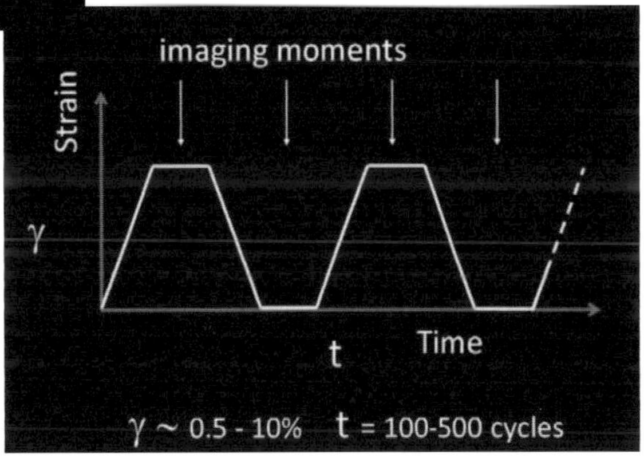

Strobed images

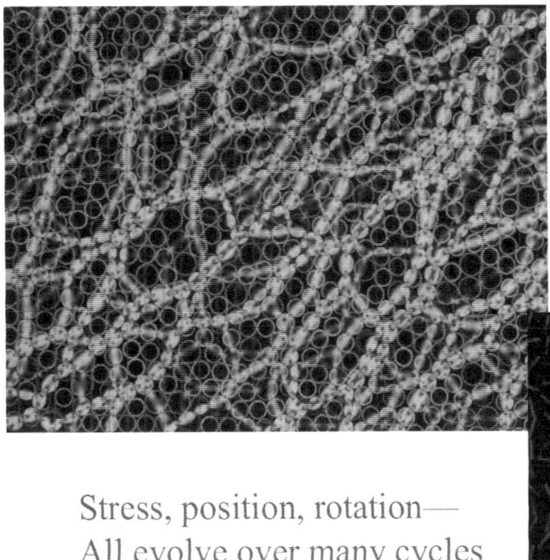

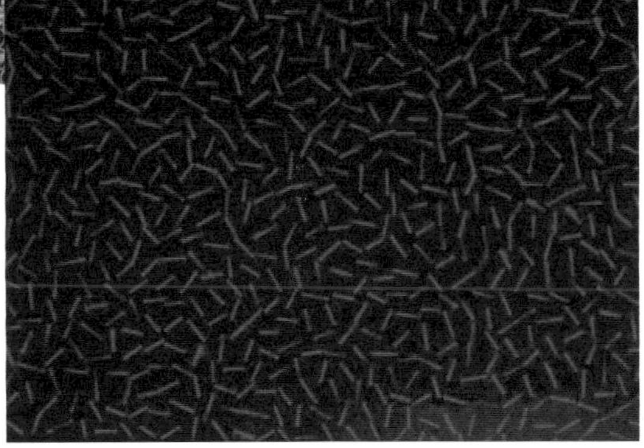

Stress, position, rotation—
All evolve over many cycles

First—stresses vs. unidirectional strain below φ_J

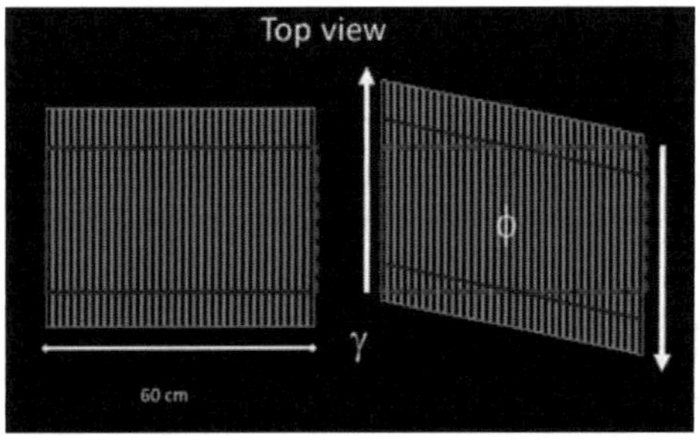

$P \approx R\gamma^2$

Define Reynolds coefficient, R

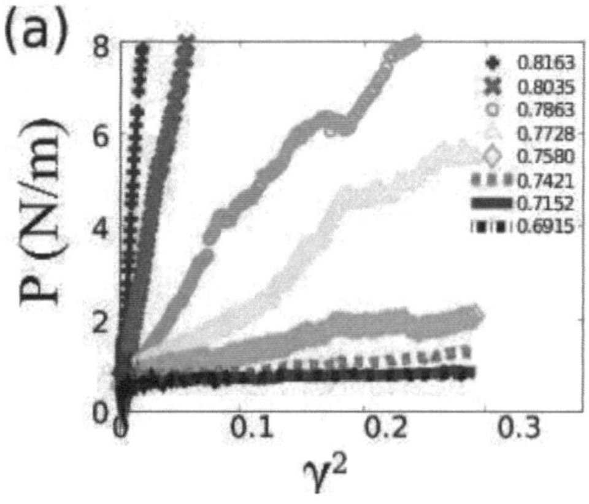

Ren et al.
PRL 110, 018302 (2013)

Shear jamming dynamics below φ_J

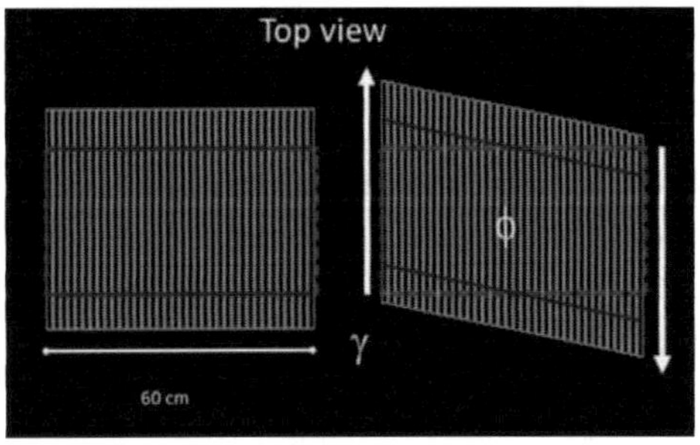

Define Reynolds coefficient, R

$P \approx R\gamma^2$

$R \sim (\varphi_J - \varphi)^3$

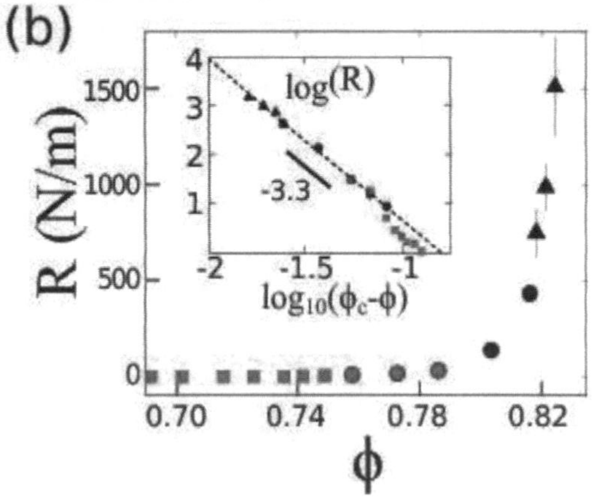

Apply asymmetric cyclic shear: note slow relaxation

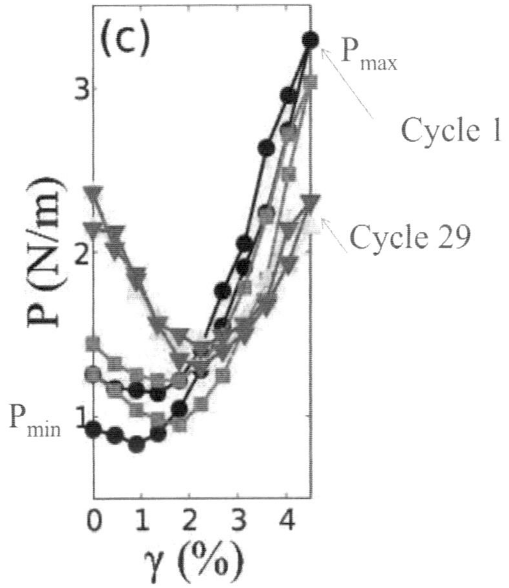

$\Delta P = P_{max} - P_{min}$

Asymmetric shear:
1) log-relaxation:
2) simple φ and γ dependence

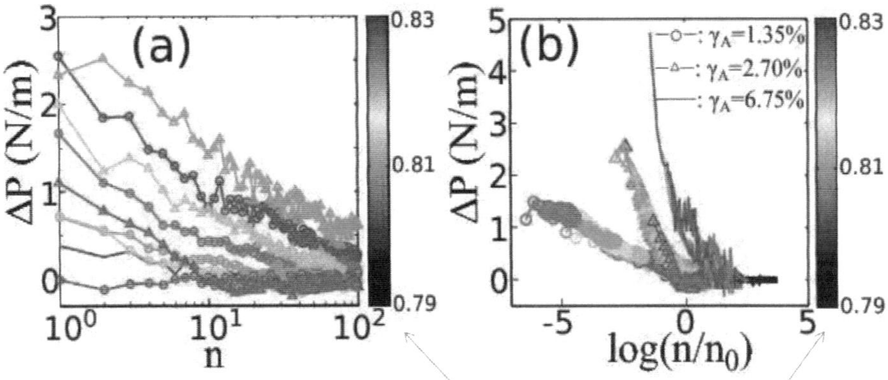

$\Delta P = -\beta \ln(n/n_o)$ Colors encode density

Universal relaxation:
consistent with activated process in a stress ensemble

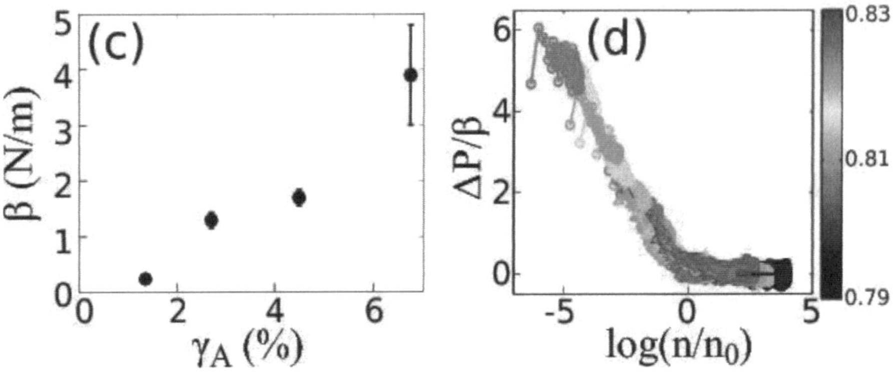

How does shear jamming affect macroscopic flows?

One frame, showing jam and force chain arch

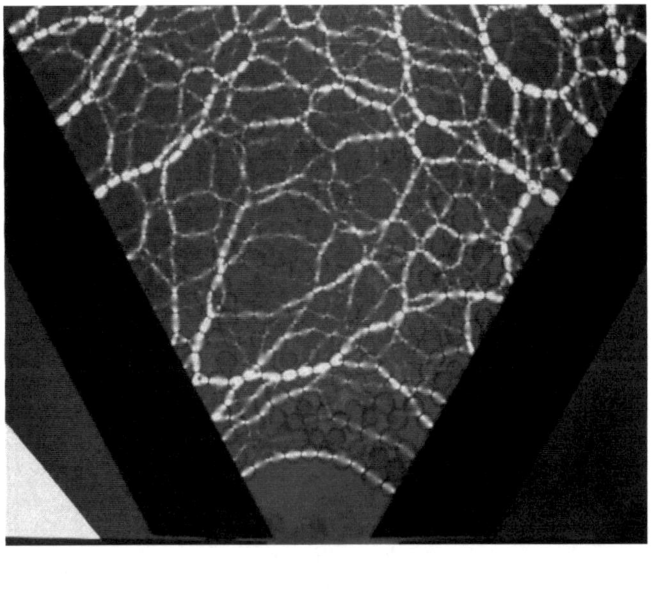

Simultaneous Density and Pressure leading to jamming event

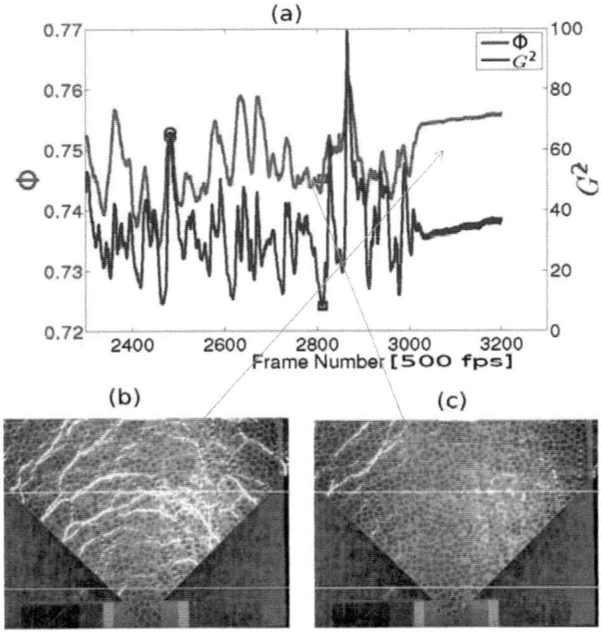

<- flow/jamming near shear jamming

Fully 3D laser scanning experiments now operational

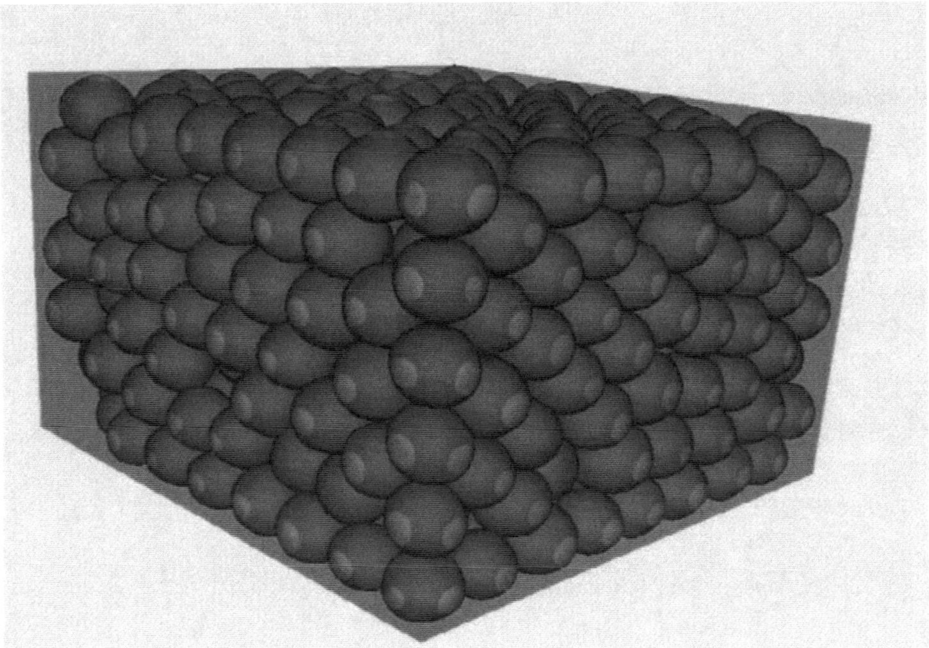

Let's end with a 'bang', or at least a thump

- Dynamics of granular impact

Past work by: Durian, Debruyn, Swinney, Goldman

See Clark and BB, Phys. Rev. Lett. **109**, 238302 (2012)

2D Granular Impact Dynamics with Photoelastic Particles:
Interesting questions of coarse-graining and scales

How is energy lost? What is the force/deceleration during the impact process? How does the material respond?

Using photoelastic particles—reveals effect of persistent force networks

Schematic of Experiment

1Mp frames at rates to ~50,000fps
reduced resolution to much higher rates

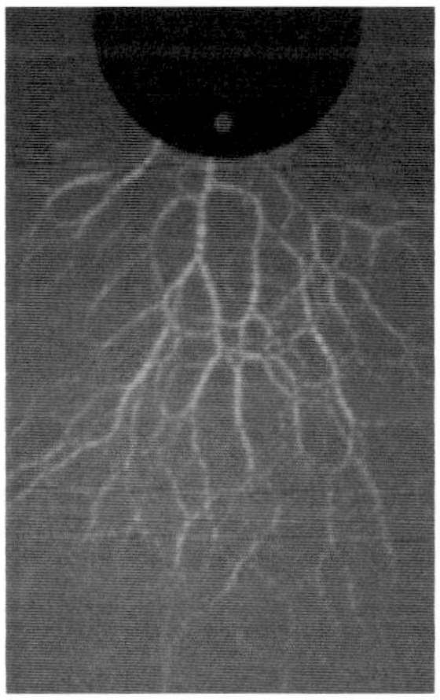

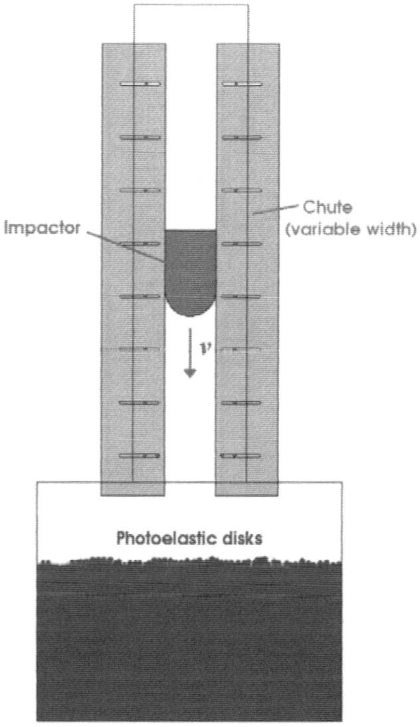

High speed photoelastic video of impact

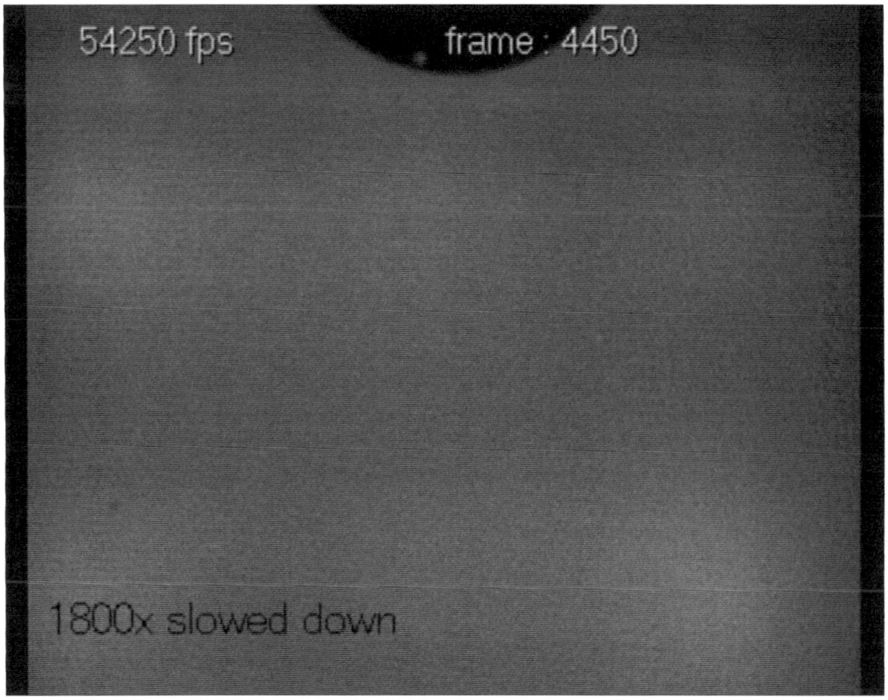

Summing up
Slow expts

- Under shear: jamming occurs for frictional particles where it it was not expected—considerable relevance for practical flows
- Fragile and fully jammed states/networks

- Key question: how to characterize states
- Strain and density are not best variables
- f_{NR} is a step in the right direction

- Fragile and shear jammed states need a network description
- Novel dynamics under cyclic shear→ stress-activated process

Open Questions

- Does the shear jamming process effectively control dense flows?

- How do other particle properties affect shear jamming? (shape, cohesion, dimension….)

- From cyclic shear experiments—sampling highly important—how do we include 'protocol'

- What are best variables for describing all aspects of jamming and flow?

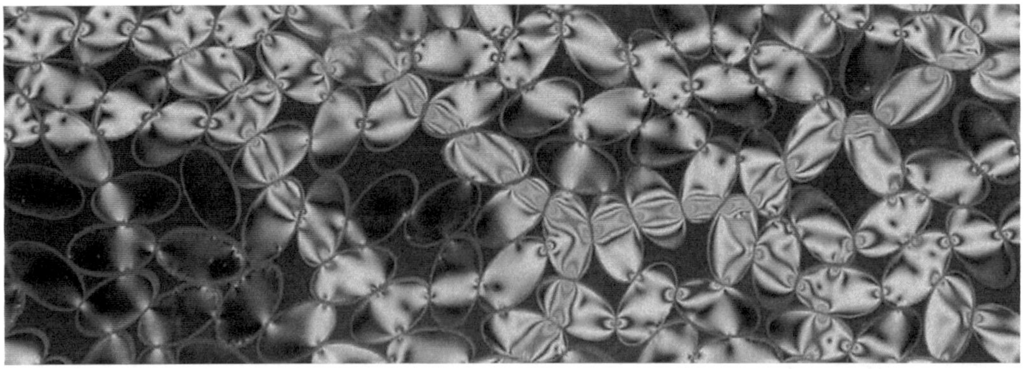

Focus of Some Scale-Up issues for Gas-Solid Flows with Group B Particles

Amblard, B., IFP Energies Nouvelles

Originally presented on: 4/30/2014 11:15:00 - 11:40:00

Chemical Looping Combustion is an oxy-combustion like technology where oxygen-carrying particles are used to supply oxygen for combustion. The process uses dense group B metal oxide as an oxygen carrier to transfer oxygen from an air reactor to a fuel reactor in a circulating fluidized bed. Process development and scale-up issues require good understanding of particle flow. This paper discusses two important aspects that were investigated in detail. The first one deals with riser flow. Experiments were conducted in a large cold flow model with two monodispersed group B solids having similar density and size, but different shape factors (sand and glass beads). For a given condition, pressure drop in the riser roughly decreases by 50% with the spherical particles. This work, therefore suggests that particle sphericity has a strong impact on the riser hydrodynamic that is not presently accounted by models. The second aspect relates to attrition. Existing knowledge of attrition is mostly based on FCC background using standard tests such as PSRI Jet-Cup apparatus. For group B powders, we found useful to define a new attrition index and also to evaluate with CFD the jet cup internal circulation rate of CLC particles compared to FCC particles.

Building a Bridge Between the Particle Scale and Bulk Properties

By Kerry Johanson

FPST Conference

April 30, 2014

Contact Information:

Dr. Kerry Johanson
Material Flow Solutions, Inc.
3536 NW 97th Blvd
Gainesville, FL 32606

352-303-9123

MatFlowSol@bellsouth.net

www.MatFlowSol.com

Material Flow Solutions Inc. - We solve problems - www.matflowsol.com - matflowsol@bellsouth.net

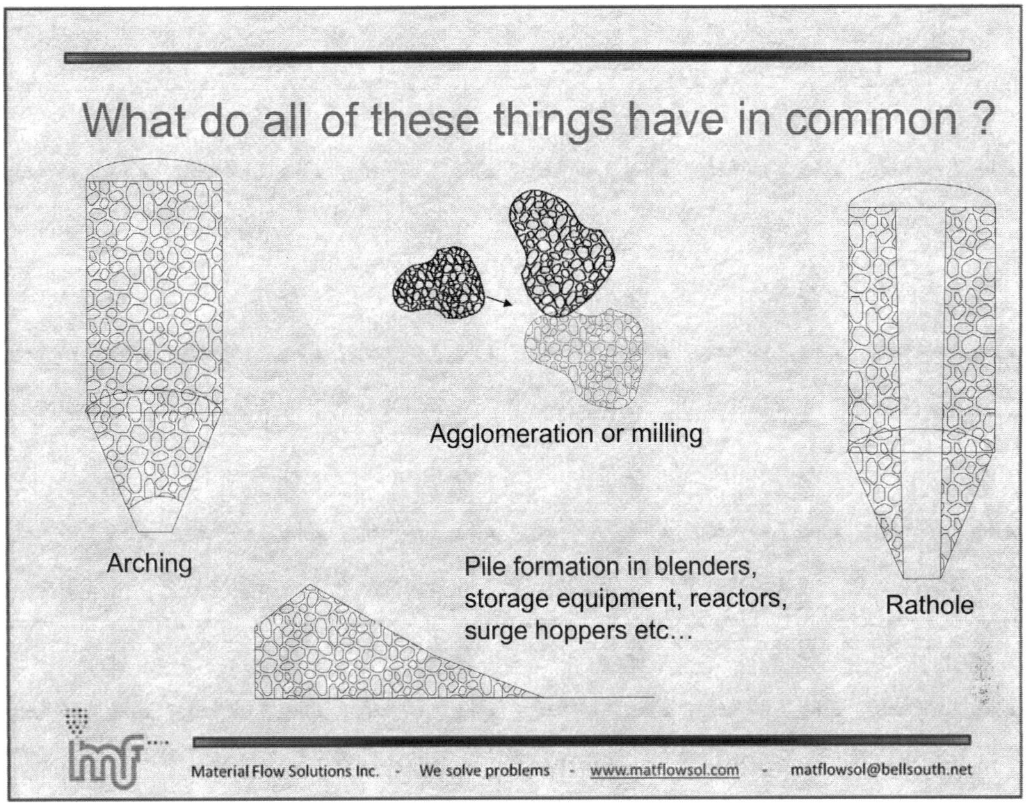
What do all of these things have in common ?
Agglomeration or milling
Arching
Pile formation in blenders, storage equipment, reactors, surge hoppers etc…
Rathole
Material Flow Solutions Inc. - We solve problems - www.matflowsol.com - matflowsol@bellsouth.net

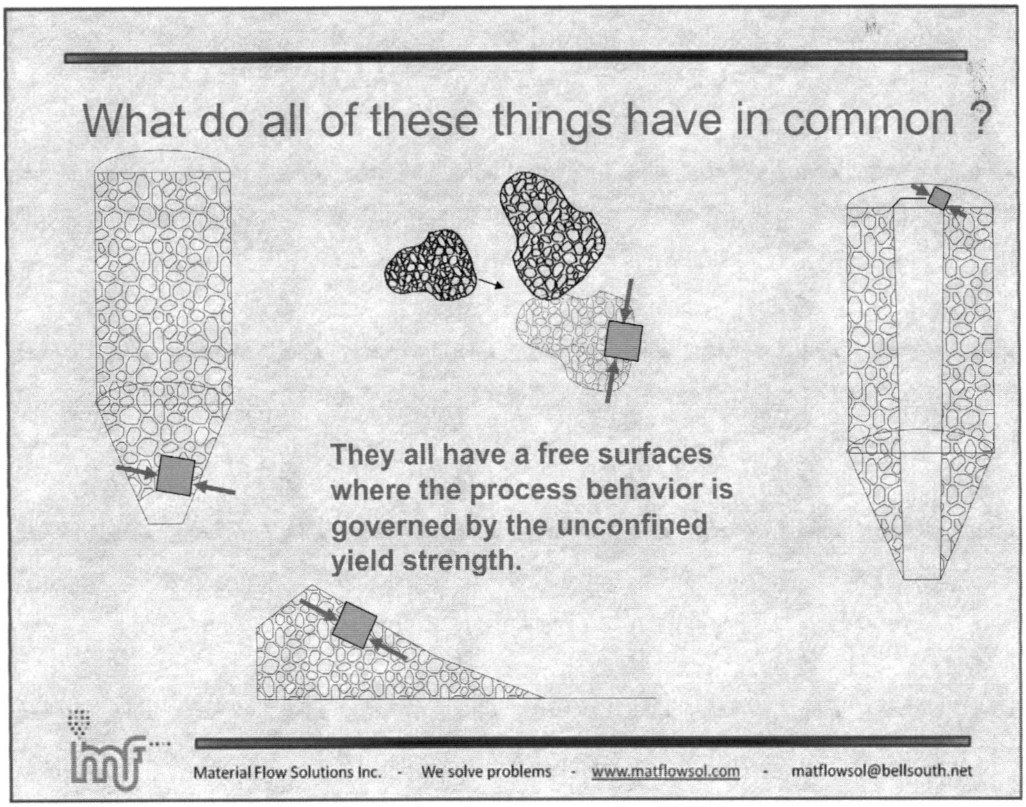
What do all of these things have in common ?
They all have a free surfaces where the process behavior is governed by the unconfined yield strength.
Material Flow Solutions Inc. - We solve problems - www.matflowsol.com - matflowsol@bellsouth.net

Understanding what influences strength will help us predict:

- Arching
- Rathole
- Blending
- **Agglomeration**
- **Milling**
- Segregation
- Fluidization
- Caking
- Product Quality

To model Milling or Attrition You need to understand two things

Strength of bulk material

- Capillary effects
- Van der Waal
- Elastic plastic effects
- Ploy disperse effects
- Shape effects

Structure

Interaction mode with surfaces

- Fracture
- Fatigue
- Attrition
- Velocity
- Interaction angle

Interaction

- Use gas pressure to drive particles to target plate
- Measure the velocity of particle
- Measure the breakage size distribution resulting from impact
- Determine mode of breakage

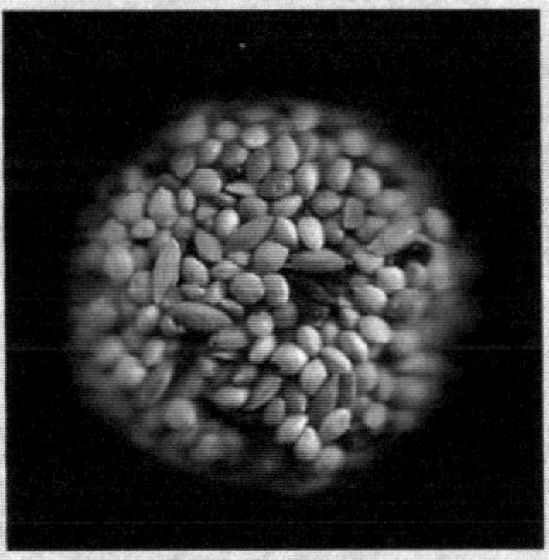

Impact particle on plate

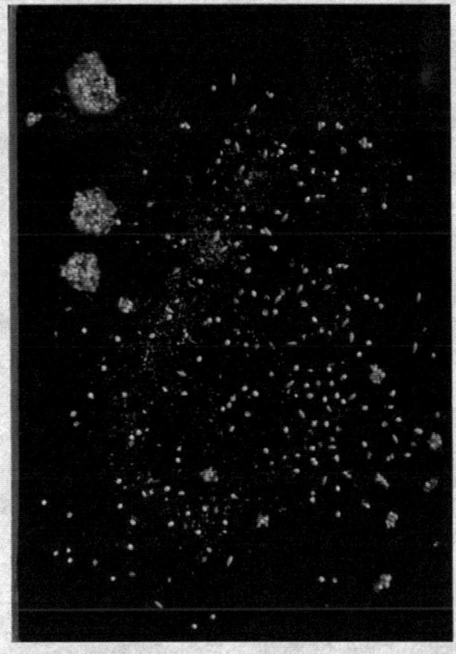

Material Flow Solutions Inc. - We solve problems - www.matflowsol.com - matflowsol@bellsouth.net

Glancing blow results in increased attrition

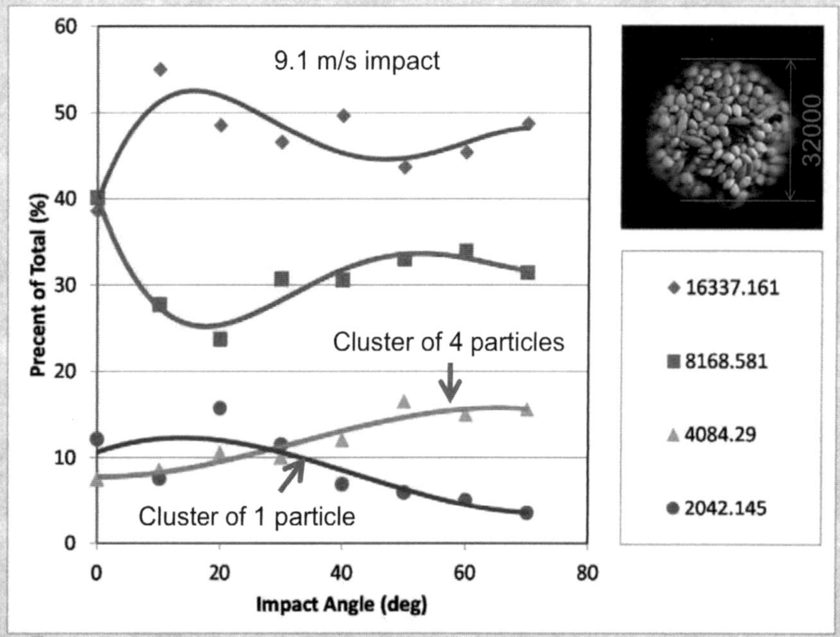

Material Flow Solutions Inc. - We solve problems - www.matflowsol.com - matflowsol@bellsouth.net

$$\frac{1}{E^*} = \frac{1-\nu_1^2}{E_1} + \frac{1-\nu_2^2}{E_2}$$

Average Elastic Modulus E*

$$\frac{1}{R} = \frac{1}{R_1} + \frac{1}{R_2}$$

Effective particle radius R

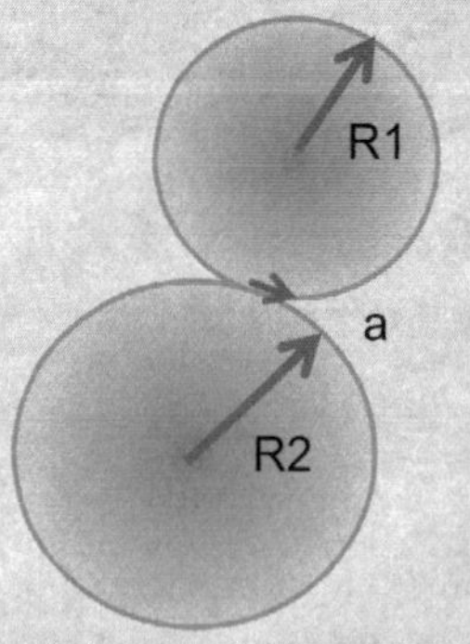

$$p_0 = \frac{2}{\pi} E^* \left(\frac{d}{R}\right)^{1/2}$$

Max stress

$$p = p_0 \left(1 - \frac{r^2}{a^2}\right)^{1/2}$$

Stress distribution

a is the contact radius
Max shear at 0.49 a depth in sphere

Material Flow Solutions Inc. - We solve problems - www.matflowsol.com - matflowsol@bellsouth.net

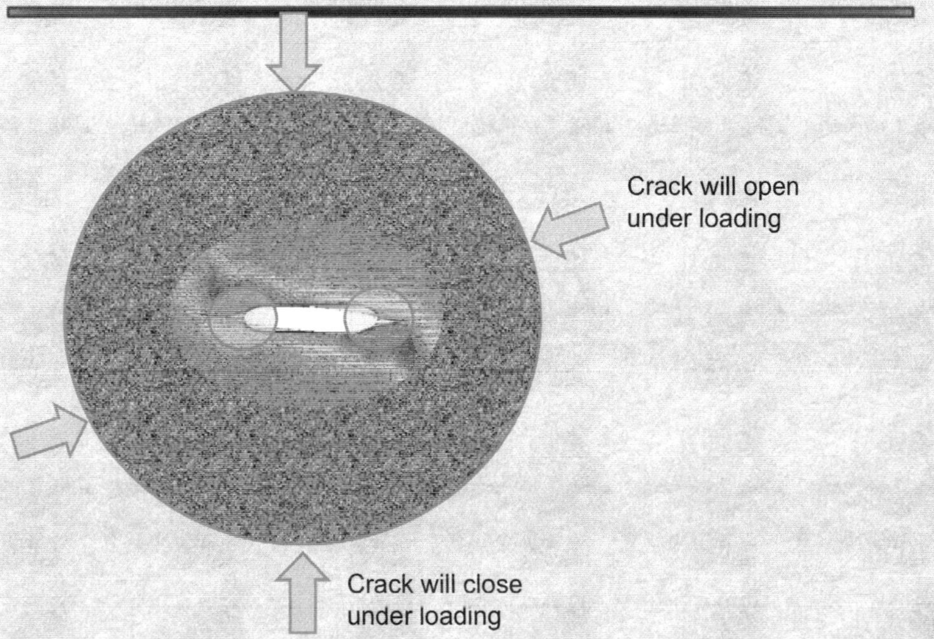

Material Flow Solutions Inc. - We solve problems - www.matflowsol.com - matflowsol@bellsouth.net

The major principle stress σ_1 increases and moves ahead of contact point as shear contact force Qs increases. stress τ_{max} generate several concentrated points. High shear points moves to the surface and results in small crack formation on surface (attrition).

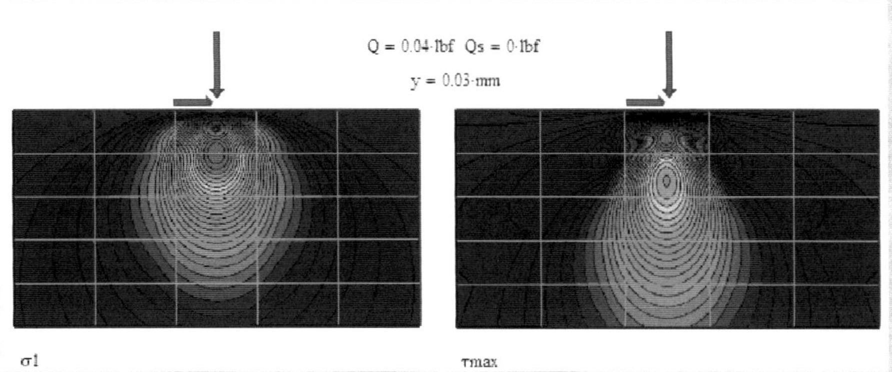

Q = normal force
Qs = shear force

Thus a glancing blow results in increased attrition

Material Flow Solutions Inc. - We solve problems - www.matflowsol.com - matflowsol@bellsouth.net

Test Procedure Details

Strength Test

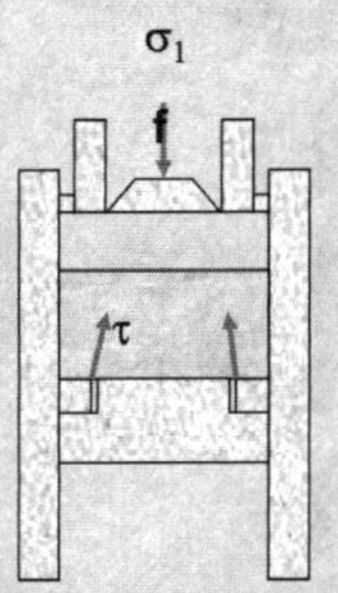

- Consolidate material in cell to a pressure of $\sigma 1$.
- Raise upper shroud.
- Lower bottom piston.
- Drive inner piston through material.
- Measure shear stress
- Compute yield strength

Note: Material can heated prior to load application and held under consolidation load for a storage time prior to failure.

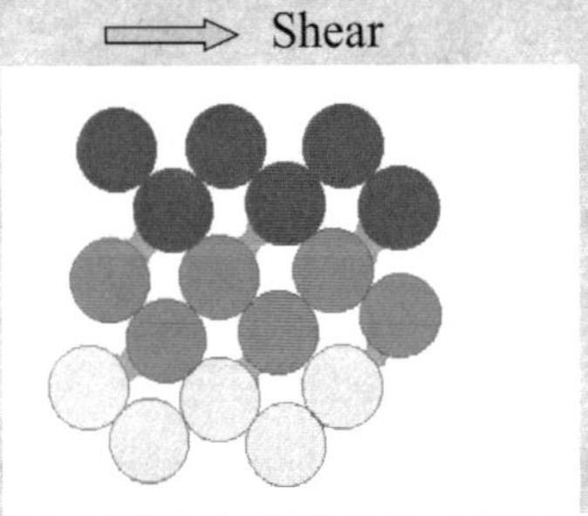

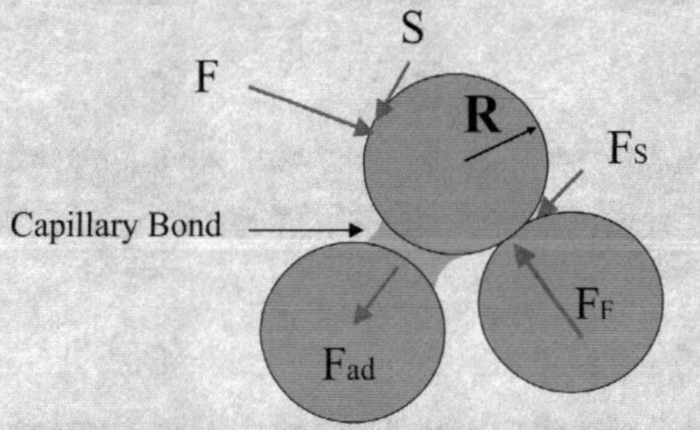

$$E_{mech} = \tau \cdot A_{sl} \cdot 2 \cdot R$$

Mechanical work to move the entire array of particles past the next adjacent particle

$$W_{per\_step} = W_1 \cdot N \cdot A_{sl} \cdot \frac{Cn}{2} \cdot n_{layer}$$

Work to separate the adhesive bridges between the particles

W_1 = work to separate one bridge
N =number of particles in shear area
n_{layer} = number of shearing layers
Cn = coordination number

INCORPORATION OF FRICTION FORCE

$$W_{mech} = E_{br} + E_{fr}$$

$$f_c = \frac{4\gamma k_r n_l p \cos\theta}{\sqrt[3]{k_h} R \cos\phi} \sqrt{\frac{C n_c \rho_{sol}}{3\rho_{liq}}} + \boxed{\frac{\mu A_h n_c n_l p \sqrt[3]{k_r^2}}{12\pi R H^2 rf \cos\phi}}$$

γ - surface tension
ρ_{liq} – density of binder
n_c – coordination number
p – probability of rupture
k_h – hexagonal packing factor
rf – roughness factor
θ - contact angle
μ - friction coefficient
n_l – number of shear layers
k_r – random packing factor
H – separation distance
ϕ - angle of internal friction
ρ_{sol} – density of particles
A_h – Hamaker constant
C - Liquid concentration
R - Particle radius

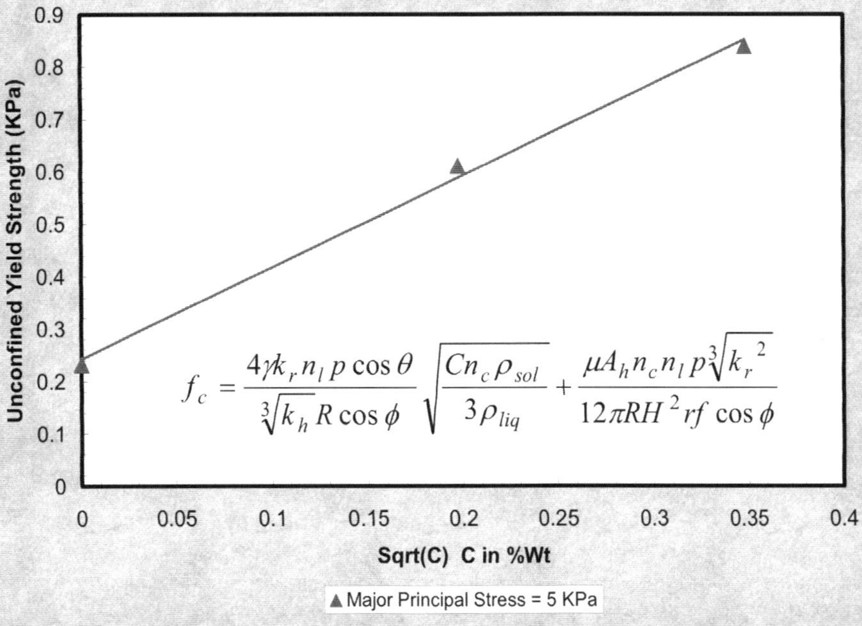

Material Flow Solutions Inc. - We solve problems - www.matflowsol.com - matflowsol@bellsouth.net

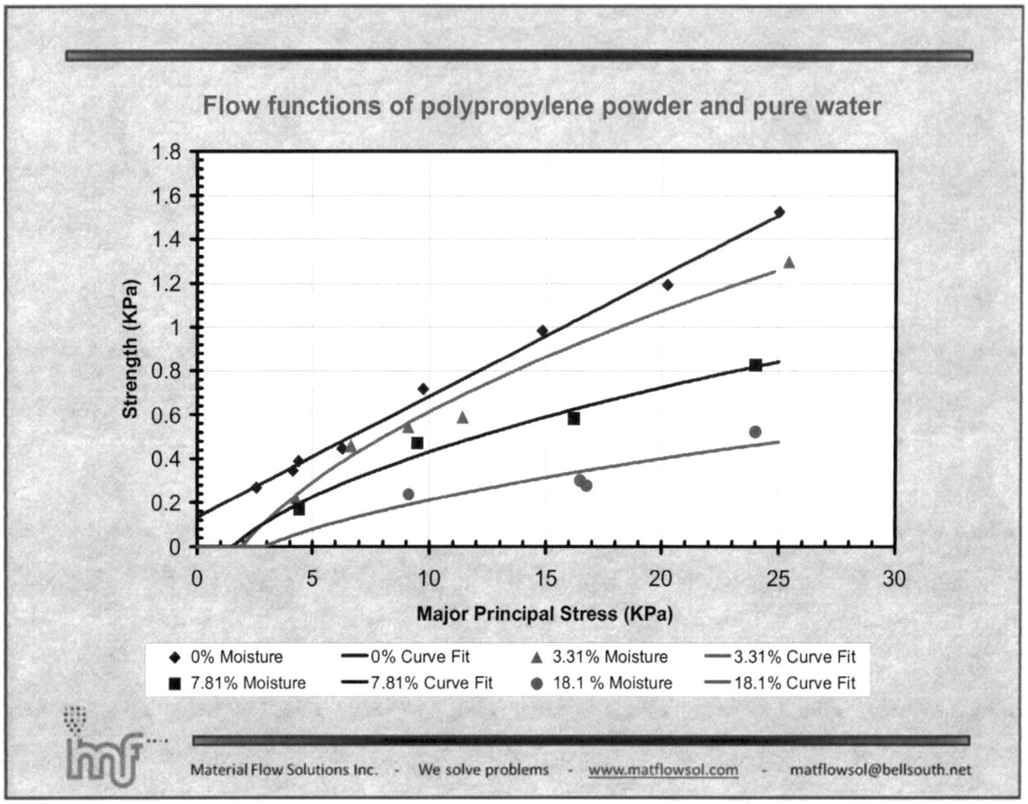
Flow functions of polypropylene powder and pure water
Strength (KPa)
1.8
1.6
1.4
1.2
1
0.8
0.6
0.4
0.2
0
0
5
10
15
20
25
30
Major Principal Stress (KPa)
0% Moisture
0% Curve Fit
3.31% Moisture
3.31% Curve Fit
7.81% Moisture
7.81% Curve Fit
18.1 % Moisture
18.1% Curve Fit
Material Flow Solutions Inc. - We solve problems - www.matflowsol.com - matflowsol@bellsouth.net

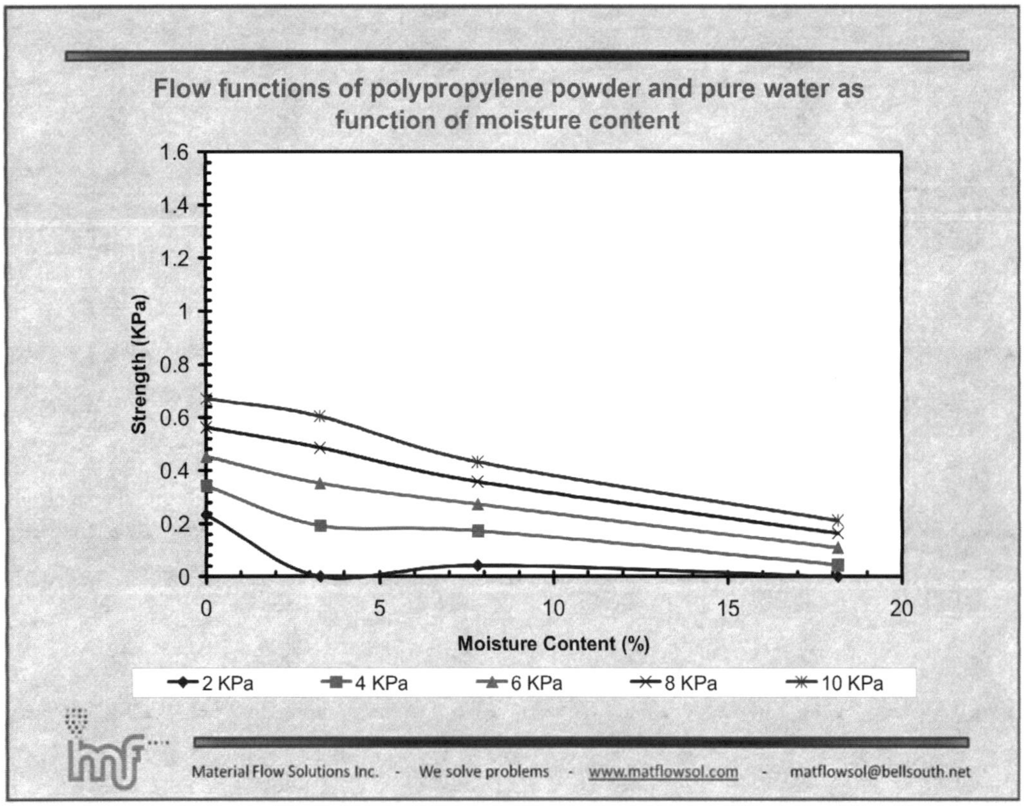
Flow functions of polypropylene powder and pure water as function of moisture content
Strength (KPa)
1.6
1.4
1.2
1
0.8
0.6
0.4
0.2
0
0
5
10
15
20
Moisture Content (%)
2 KPa
4 KPa
6 KPa
8 KPa
10 KPa
Material Flow Solutions Inc. - We solve problems - www.matflowsol.com - matflowsol@bellsouth.net

Flow functions of polypropylene powder and mixture of 70% alcohol and 30% water

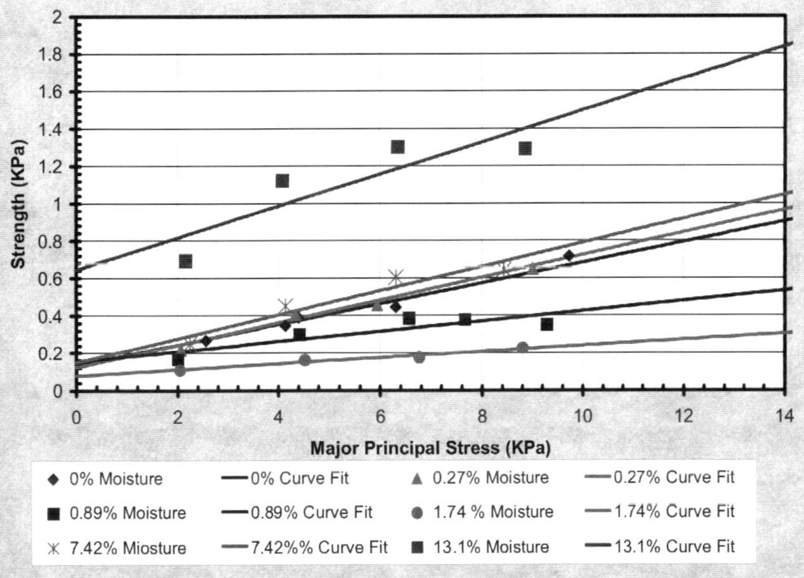

Material Flow Solutions Inc. - We solve problems - www.matflowsol.com - matflowsol@bellsouth.net

Flow functions of polypropylene powder and mixture of 70% alcohol and 30% water as a function of moisture content

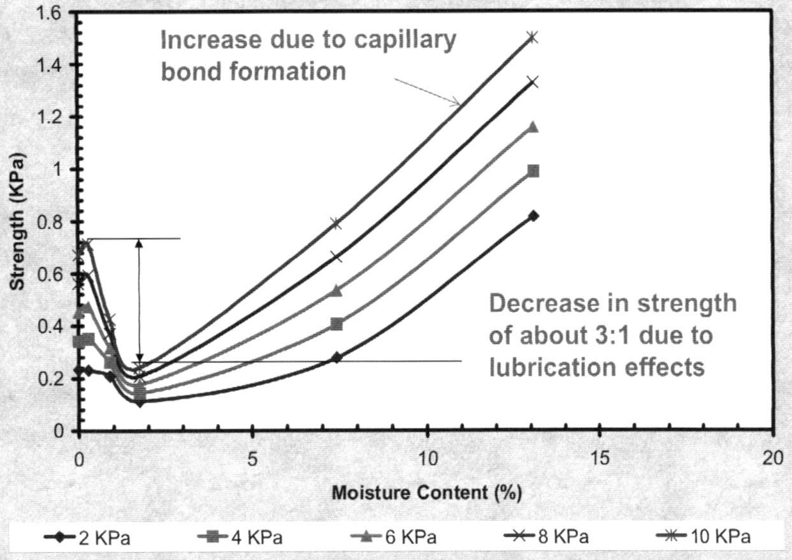

Material Flow Solutions Inc. - We solve problems - www.matflowsol.com - matflowsol@bellsouth.net

Model	Mechanism	Source
$fc = \frac{K_1}{Dp^2}$	Van der Waals forces	Mollerus [4]
$fc = \frac{K_2 \cdot \sqrt{C}}{Dp}$	Capillary forces	Rabinovich [2]
$fc = \frac{K_3}{Dp^{1/2}}$	Elastic fracture	Rumpf [1]
$fc = \frac{K_4}{Dp^n}$	Plastic-elastic fracture	Spetch [3]

$$fc = \frac{K_1 \cdot \sqrt{C}}{D_p} \cdot X_{capillary} + \frac{K_2}{D_p^{\ 2}} \cdot X_{van} + \frac{K_3}{D_p^{\ n}} \cdot X_{fracture}$$

Material Flow Solutions Inc. - We solve problems - www.matflowsol.com - matflowsol@bellsouth.net

Poly-disperse Material

Material Flow Solutions Inc. - We solve problems - www.matflowsol.com - matflowsol@bellsouth.net

Case 1

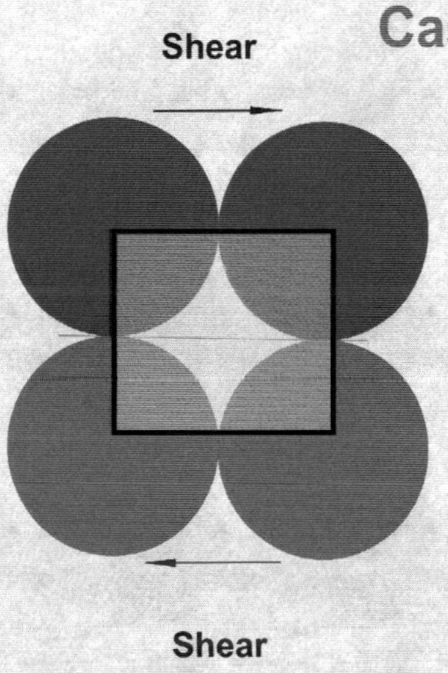

There are no small particles in the system. Shear takes place between adjacent large particles. The strength of this structure is found from the strength values for mono-dispersed systems for large particles.

$$fc = \frac{K}{Dp_A}$$

Case 2

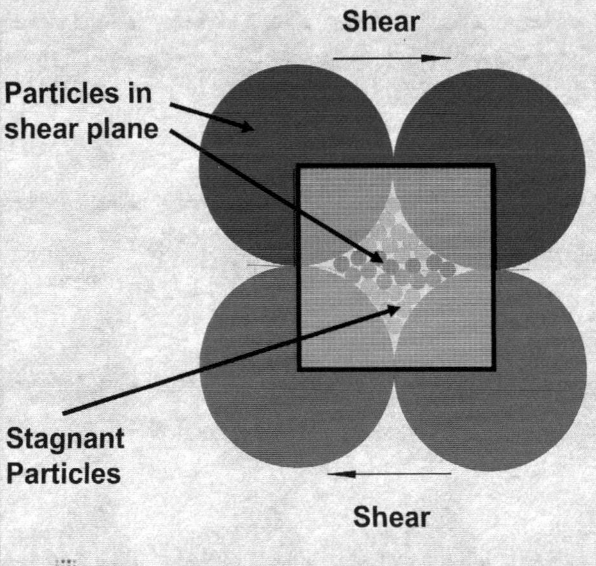

The voids are filled with small particles. Shear takes place between both large and small particles. The strength of this structure is found from the combined strength of both small and large particles.

Material Flow Solutions Inc. - We solve problems - www.matflowsol.com - matflowsol@bellsouth.net

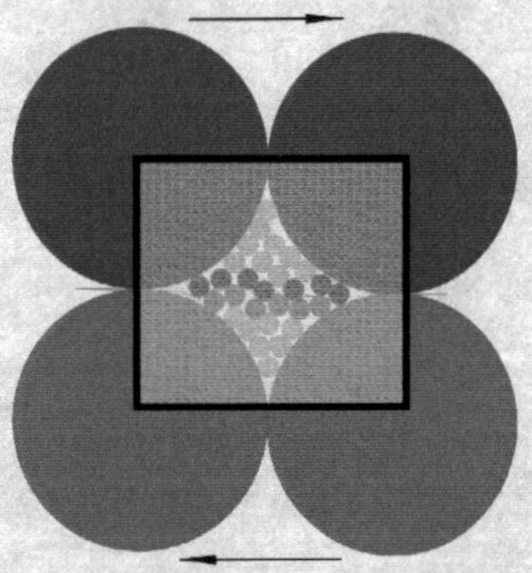

$$fc = \left[\frac{K}{Dp_B}\cdot\left(\frac{4}{\pi}-1\right)\cdot\left(\frac{Dp_A}{Dp_B}\right)^2\cdot PF_B\cdot fr_{void}\right] + \frac{K}{Dp_A}$$

Number of small particle contacts per unit area of shear. Shear area defined by the distance between large particles.

Dp_A is the particle size of the large particles
Dp_B is the particle size of the small particles
PF_B is the packing factor of the small particle matrix
fr_{void} is the fraction of total small particles that are involved in shear
K is a function of surface tension and contact angle and moisture

Material Flow Solutions Inc. - We solve problems - www.matflowsol.com - matflowsol@bellsouth.net

Case 3

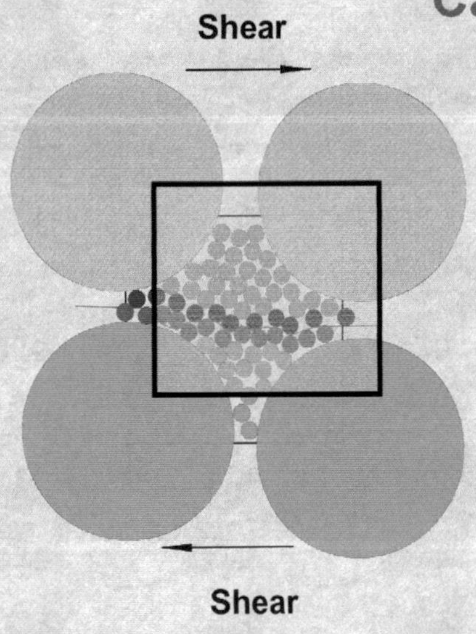

There are an excess of small particles in the system. Shear takes place between adjacent small particles. The strength of this structure is found from the strength values for mono-dispersed systems for small particles.

$$fc = \frac{K}{Dp_B}$$

Material Flow Solutions Inc. - We solve problems - www.matflowsol.com - matflowsol@bellsouth.net

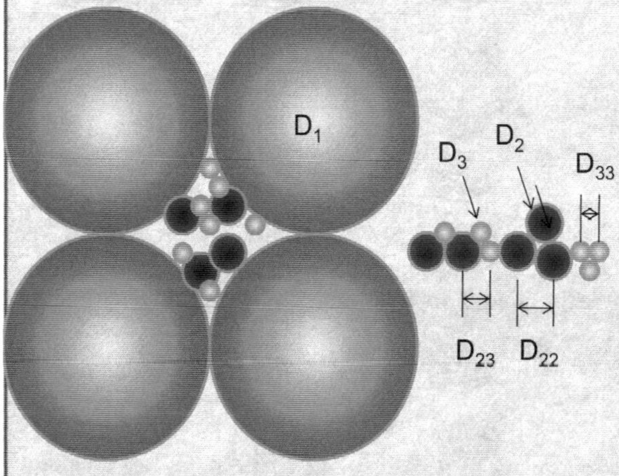

Assume the largest particle creates the coarse matrix where finer particle fit within the interstitial pores. The number of particles in these pores depends on the average particle size. This average particle size can be found from the equation below and then used to replace Dp_B in the model above to predict yield strength.

$$Dp_{avg} = Dp_{22} \cdot \left(f_2^3 + f_2^2 \cdot f_3\right) + 2 \cdot Dp_{23} \cdot \left(f_3^2 \cdot f_2 + f_2^2 \cdot f_3\right) + Dp_{33} \cdot \left(f_3^3 + f_3^2 \cdot f_2\right)$$

f_2 is the fraction of #2 particles in the pores
f_3 is the fraction of #3 particles in the pores

Material Flow Solutions Inc. - We solve problems - www.matflowsol.com - matflowsol@bellsouth.net

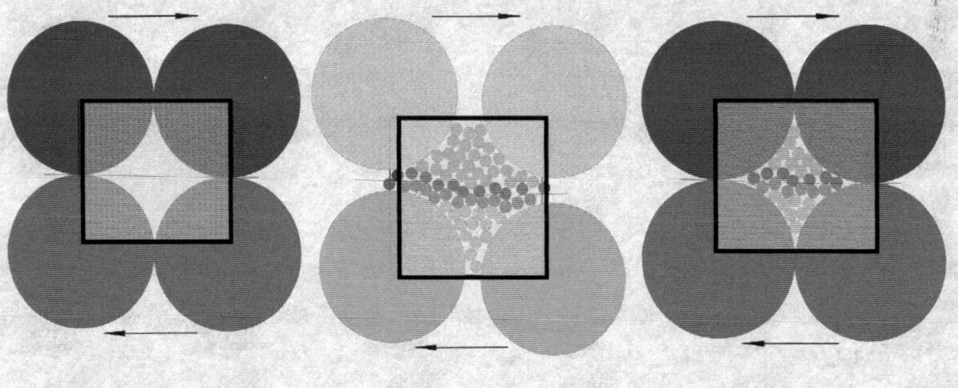

The bulk material is made up of a combination of these type of structures. So if we add the strength effects of all of these structures then we can estimate the overall strength of the mixture.

Material Flow Solutions Inc. - We solve problems - www.matflowsol.com - matflowsol@bellsouth.net

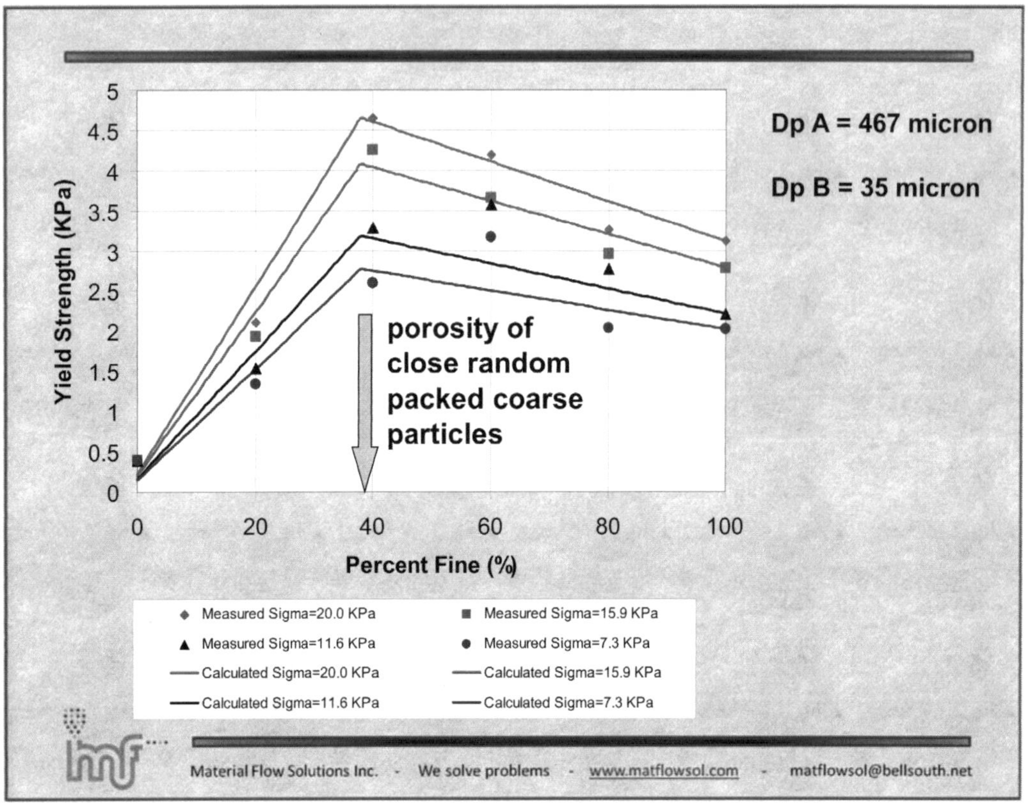
Dp A = 467 micron
Dp B = 35 micron
Yield Strength (KPa)
Percent Fine (%)
porosity of close random packed coarse particles
Measured Sigma=20.0 KPa
Measured Sigma=15.9 KPa
Measured Sigma=11.6 KPa
Measured Sigma=7.3 KPa
Calculated Sigma=20.0 KPa
Calculated Sigma=15.9 KPa
Calculated Sigma=11.6 KPa
Calculated Sigma=7.3 KPa
Material Flow Solutions Inc. - We solve problems - www.matflowsol.com - matflowsol@bellsouth.net

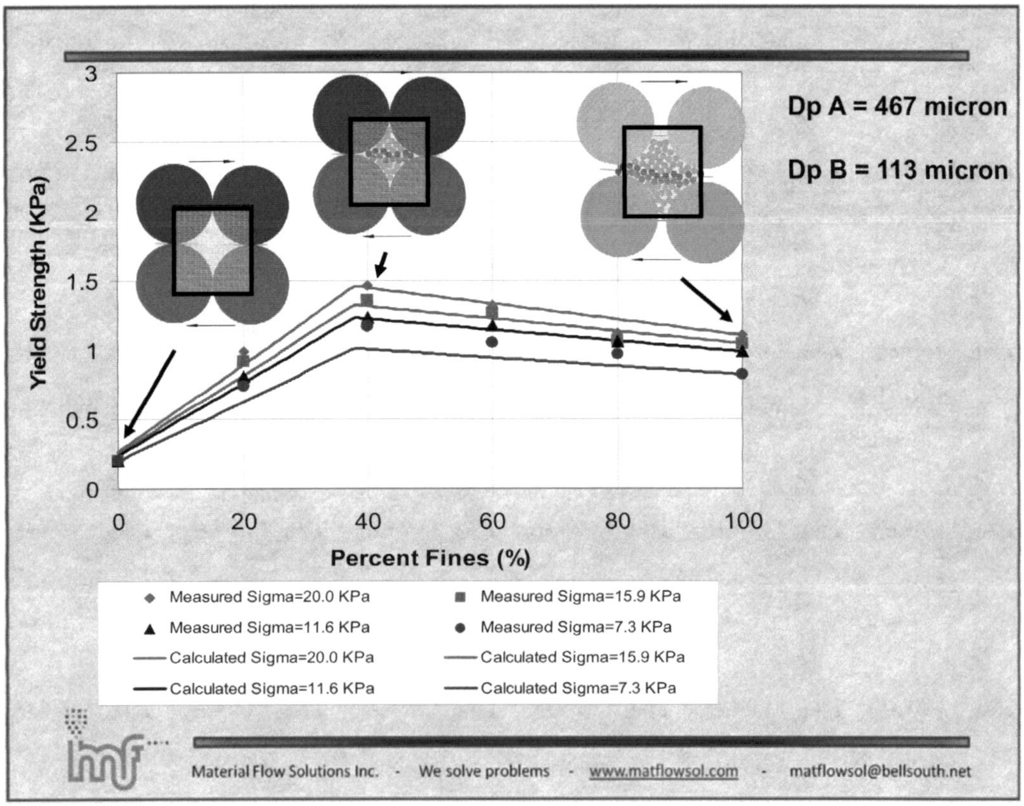
Dp A = 467 micron
Dp B = 113 micron
Yield Strength (KPa)
Percent Fines (%)
Measured Sigma=20.0 KPa
Measured Sigma=15.9 KPa
Measured Sigma=11.6 KPa
Measured Sigma=7.3 KPa
Calculated Sigma=20.0 KPa
Calculated Sigma=15.9 KPa
Calculated Sigma=11.6 KPa
Calculated Sigma=7.3 KPa
Material Flow Solutions Inc. - We solve problems - www.matflowsol.com - matflowsol@bellsouth.net

Effect of Stress

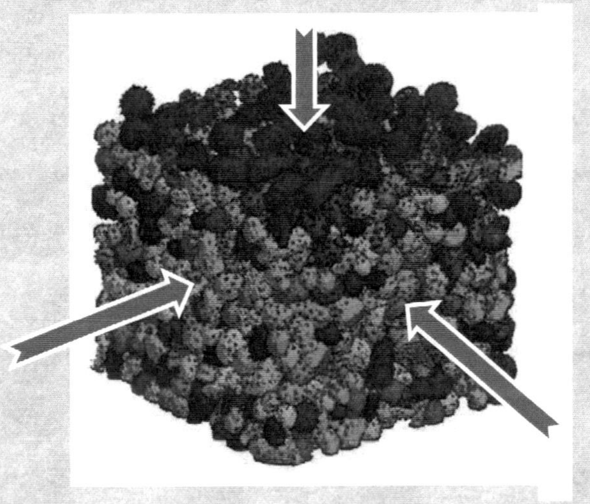

Material Flow Solutions Inc. - We solve problems - www.matflowsol.com - matflowsol@bellsouth.net

500 N Normal Force

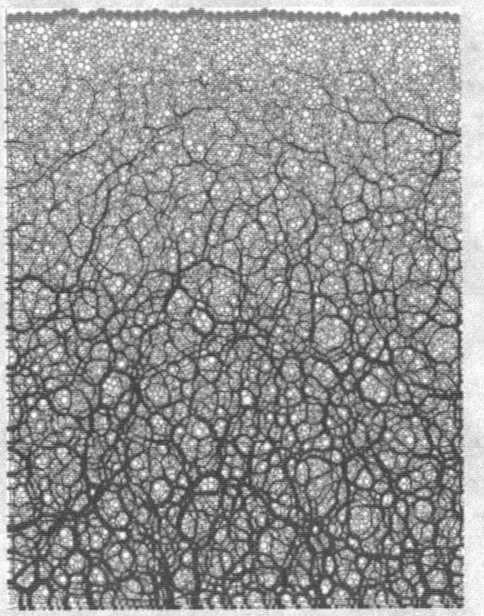

5000 N Normal Force

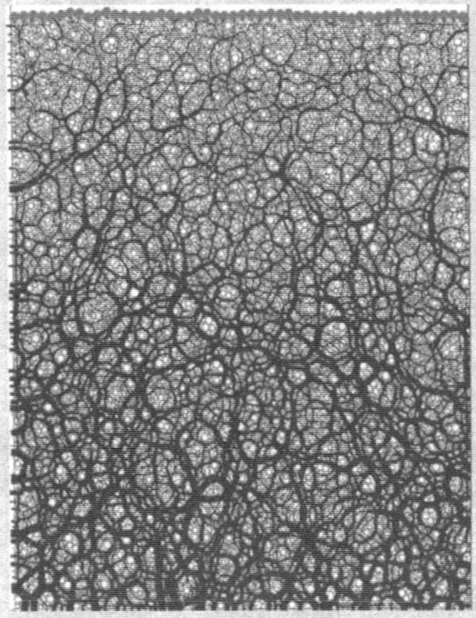

Material Flow Solutions Inc. - We solve problems - www.matflowsol.com - matflowsol@bellsouth.net

500 N Applied force

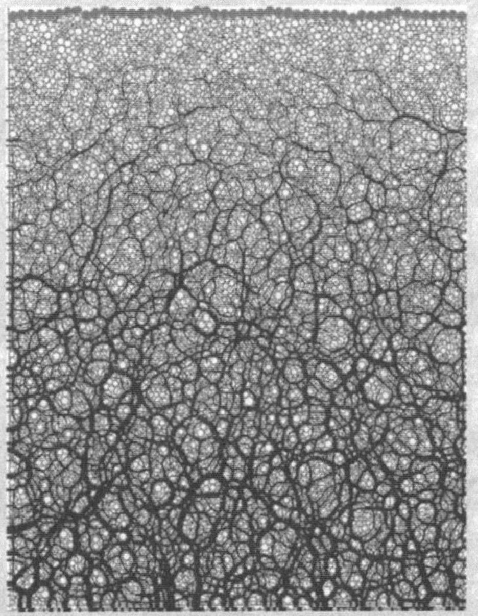

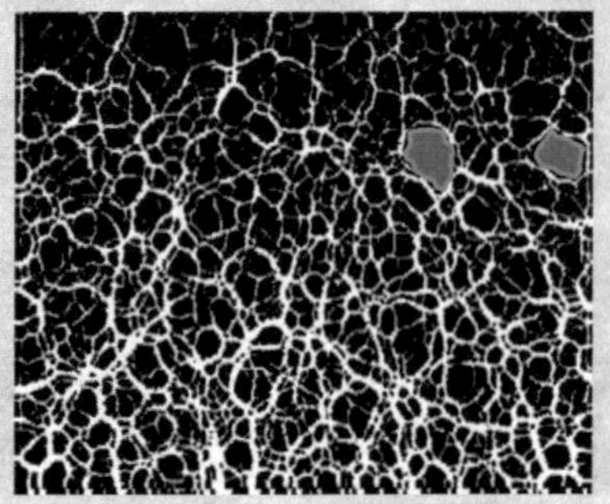

We generate the granularity of force chains by selecting a critical force and then drawing lines to connect the force chains and do an optical image analysis to generate the size of the regions between the force chains

Material Flow Solutions Inc. - We solve problems - www.matflowsol.com - matflowsol@bellsouth.net

500 N Applied force

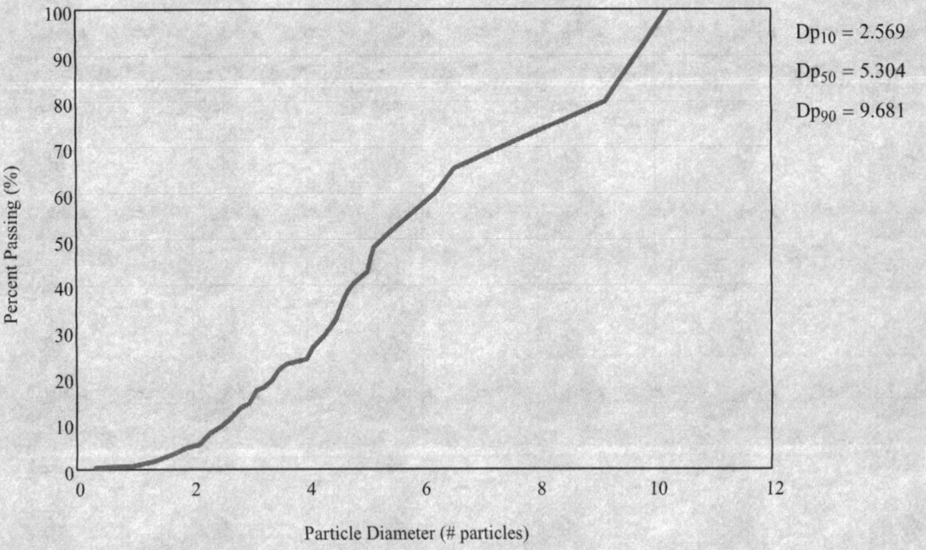

Material Flow Solutions Inc. - We solve problems - www.matflowsol.com - matflowsol@bellsouth.net

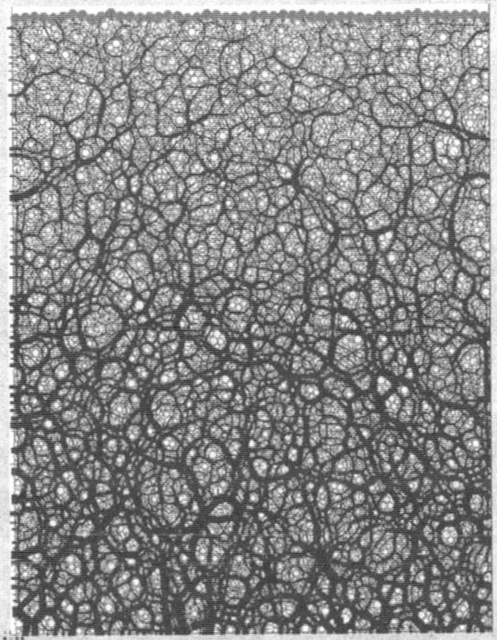

5000 N Applied Force

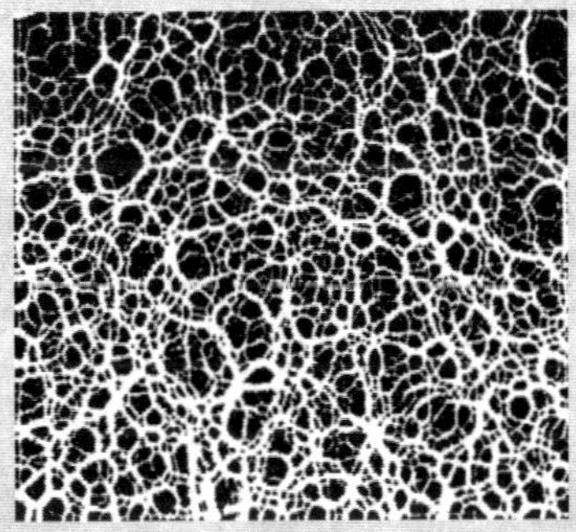

The force chains are closer together at higher pressures

Material Flow Solutions Inc. - We solve problems - www.matflowsol.com - matflowsol@bellsouth.net

5000 N Applied Force

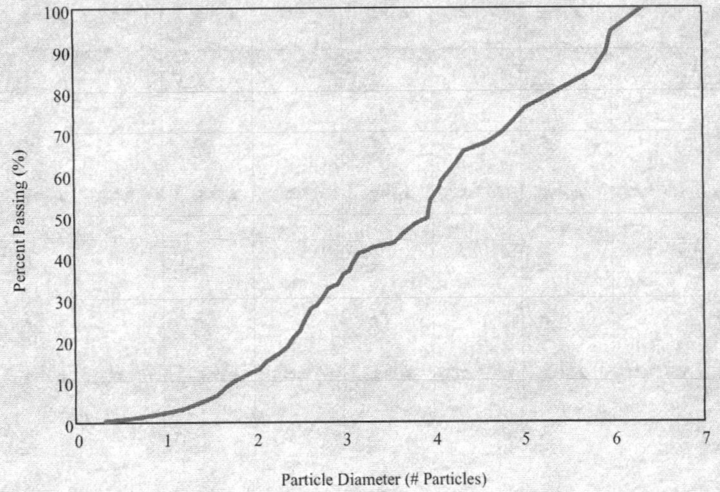

$Dp_{10} = 1.812$

$Dp_{50} = 3.943$

$Dp_{90} = 5.949$

Material Flow Solutions Inc. - We solve problems - www.matflowsol.com - matflowsol@bellsouth.net

Effect of Stress

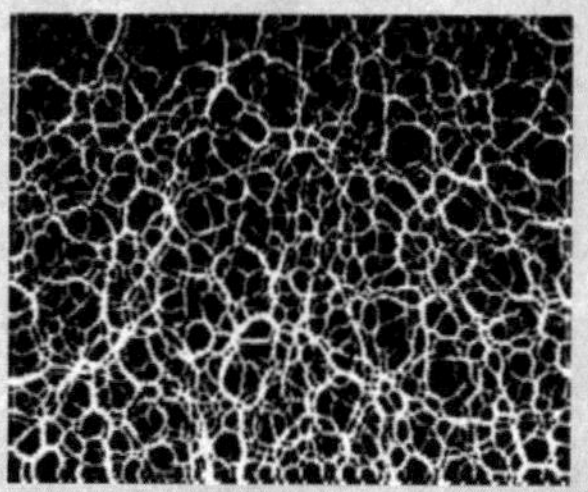

Force chain structure at 500 N

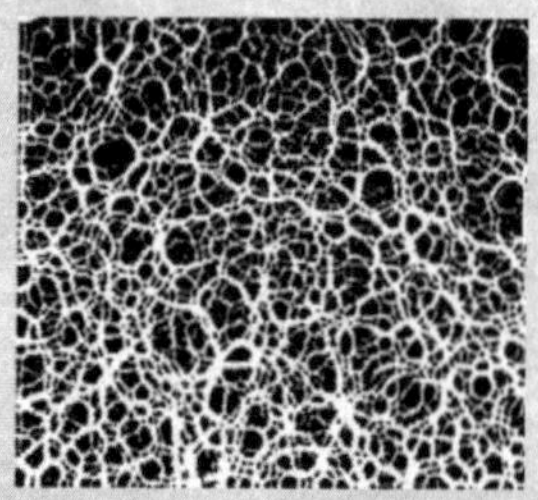

Force chain structure at 5000 N

Force chain structure shows closer force chains at higher loads

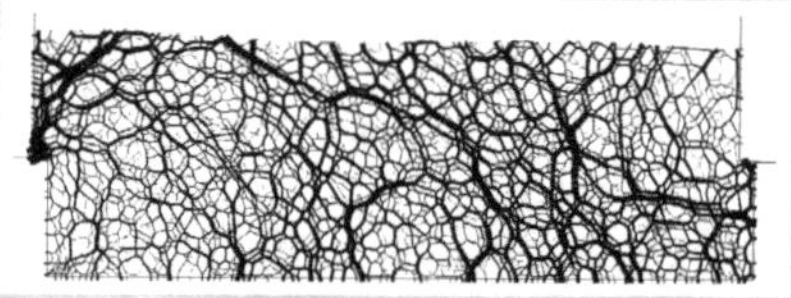

Force chain structure during shear

We believe the # of major force chain structures per unit volume helps us understand the increase of strength with increasing compaction stress.

Strength is a function of the number of force chains per unit volume

$$fc = B / (D_{space})^3$$

That force chain density depends on the stress level.

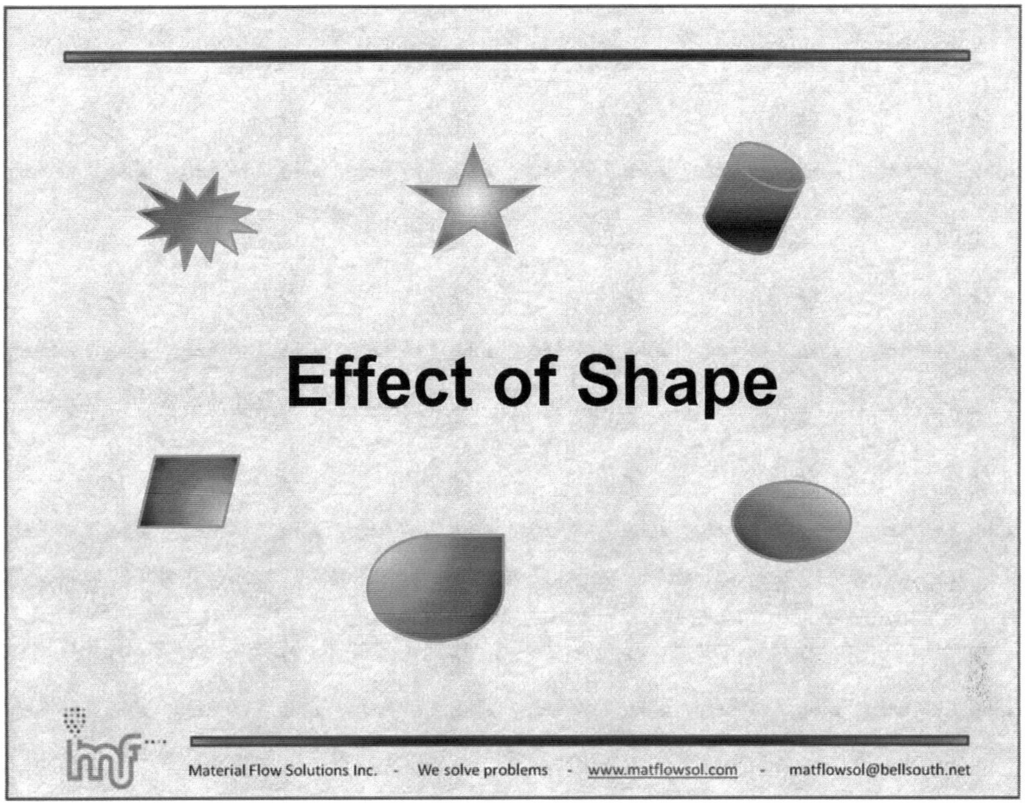

Effect of Shape
Material Flow Solutions Inc. - We solve problems - www.matflowsol.com - matflowsol@bellsouth.net

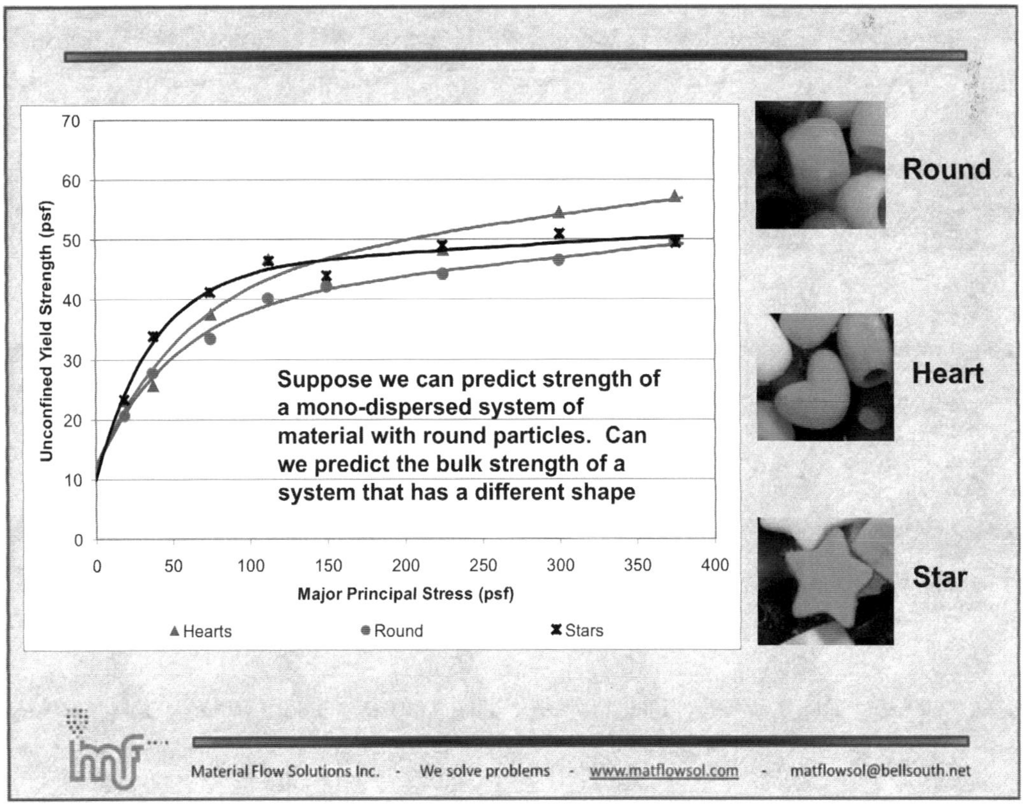

Unconfined Yield Strength (psf)
70
60
50
40
30
20
10
0
0
50
100
150
200
250
300
350
400
Major Principal Stress (psf)
Hearts
Round
Stars
Suppose we can predict strength of a mono-dispersed system of material with round particles. Can we predict the bulk strength of a system that has a different shape
Round
Heart
Star
Material Flow Solutions Inc. - We solve problems - www.matflowsol.com - matflowsol@bellsouth.net

The angle of contact between particles also effects the strength

We can count the number of contacts made by one particle against an adjacent particle. The bulk yield strength is proportional to the total number of contacts per unit volume. Thus if we count the contacts we can find a correction factor for the strength.

Material Flow Solutions Inc. - We solve problems - www.matflowsol.com - matflowsol@bellsouth.net

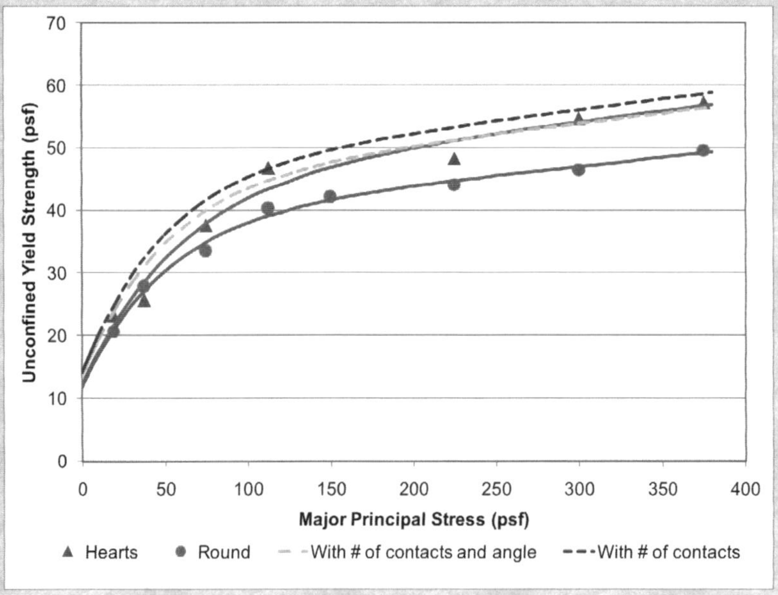

Material Flow Solutions Inc. - We solve problems - www.matflowsol.com - matflowsol@bellsouth.net

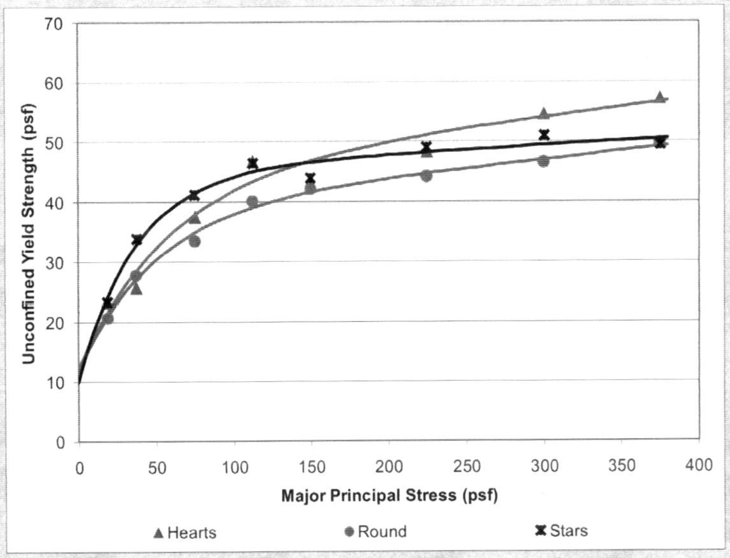

Round

Heart

Star

Material Flow Solutions Inc. - We solve problems - www.matflowsol.com - matflowsol@bellsouth.net

Bulk assembly at 0 psf

Bulk assembly at 440 psf

Notice the collapse of the structure to yield more flat to flat contacts giving rise to less contacts per particle than the heart system

Material Flow Solutions Inc. - We solve problems - www.matflowsol.com - matflowsol@bellsouth.net

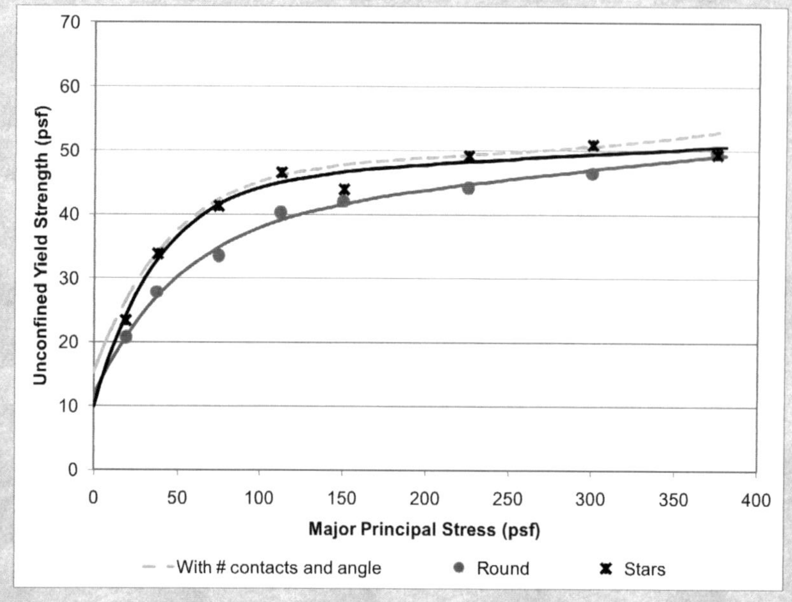

Material Flow Solutions Inc. - We solve problems - www.matflowsol.com - matflowsol@bellsouth.net

Strength can be written as a function of mechanistic terms that relate strength to some power of the particle size. Thus, know the size distribution; know the strength.

Model	Mechanism	Source
$fc = \frac{K_1}{Dp^2}$	Van der Waals forces	Mollerus [4]
$fc = \frac{K_2 \cdot \sqrt{C}}{Dp}$	Capillary forces	Rabinovich [2]
$fc = \frac{K_3}{Dp^{1/2}}$	Elastic fracture	Rumpf [1]
$fc = \frac{K_4}{Dp^n}$	Plastic-elastic fracture	Spetch [3]

$$fc = \frac{K_1 \cdot \sqrt{C}}{D_p} \cdot X_{capillary} + \frac{K_2}{D_p^{\ 2}} \cdot X_{van} + \frac{K_3}{D_p^{\ n}} \cdot X_{fracture}$$

Material Flow Solutions Inc. - We solve problems - www.matflowsol.com - matflowsol@bellsouth.net

Thank You

Material Flow Solutions Inc. - We solve problems - www.matflowsol.com - matflowsol@bellsouth.net

Modeling for Formulation Development, Scale-up and Control of Agglomeration Processes

Mort, P., Procter & Gamble Company

Originally presented on: 4/30/2014 12:15:00 - 12:45:00

Modeling can be used to aid in developing agglomerate-based product formulations and production processes. Addressing multiple objectives in the development process can be done using a set of complementary models, each with a distinct and [relatively] simple objective function. Models include formulation, process control, scale-up of unit operations and system integration. This presentation will cover some examples of useful modeling as applied to agglomerate process development.

Model Development for Particle and Tablet Coating

Wassgren, C. R., Purdue University

Originally presented on: 4/30/2014 14:00:00 - 14:25:00

Particle and tablet coating is a common unit in a variety of industries. In this talk, several recent models for predicting inter-tablet coating variability are presented including a model for continuous coating, a combined discrete element method (DEM)-compartment-population balance model for batch coating, and DEM scaling studies for batch coating.

Prediction of Particle Interactions in a Spray Dryer Using Drynetics™

Walter, M., GEA Process Engineering Inc.

Originally presented on: 4/30/2014 14:30:00 - 14:55:00

Since its recent introduction, Drynetics™ has combined Computational Fluid Dynamics with examination of the drying of a single droplet suspended in a drying gas to predict particle behavior during the spray drying process. Drynetics™ has allowed spray dryers to be more efficient, safer, smaller, of higher capacity, and with fewer operational headaches than in the past.Drynetics™ allows the examination of how individual particles dry and how different drying parameters affect the dried product. Use of the data from Drynetics™ facilitates process and product development without costly and time consuming experimental work in a pilot plant or production spray dryer. Drynetics™ also may enable remote troubleshooting of an existing spray dryer. Use of Drynetics™ may determine if a given feed will dry into hollow spheres, shriveled particles, “doughnut” shapes, etc. while providing insight into design of drying equipment and process parameters for attaining a certain type of particle.

Novel Clean Energy Technologies Utilizing High-Temperature, High-Pressure Fluidized-Bed Reactors

Raghubir Gupta, Atish Kataria, and Brian Turk
Energy Technology Division
RTI International

Frontiers in Particle Science and Technology
Oakbrook, IL
April 30, 2014

RTI International is a trade name of Research Triangle Institute.

www.rti.org

1

RTI International

Turning Knowledge Into Practice

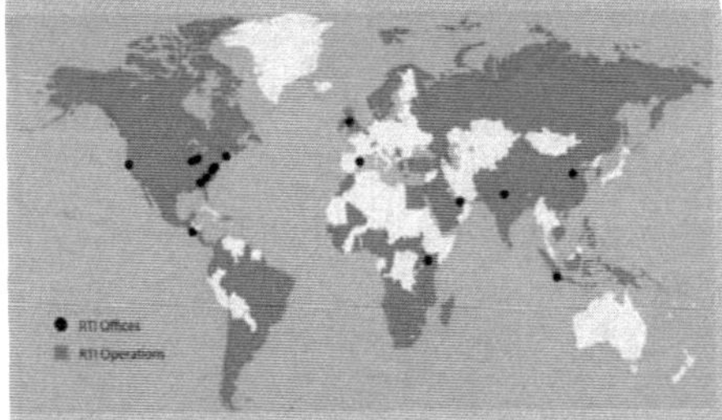

One of the world's leading research organizations

3,700 staff
Work in 75 countries

1,800+
Active RTI projects

scientific staff
Highly qualified with tremendous breadth

$780 million
Research budget

Energy Technologies

Developing advanced process technologies for energy applications by partnering with industry leaders

Biomass and Biofuels

Natural Gas

Carbon Capture

Industrial Water

Syngas

2

Integrated Technology Development

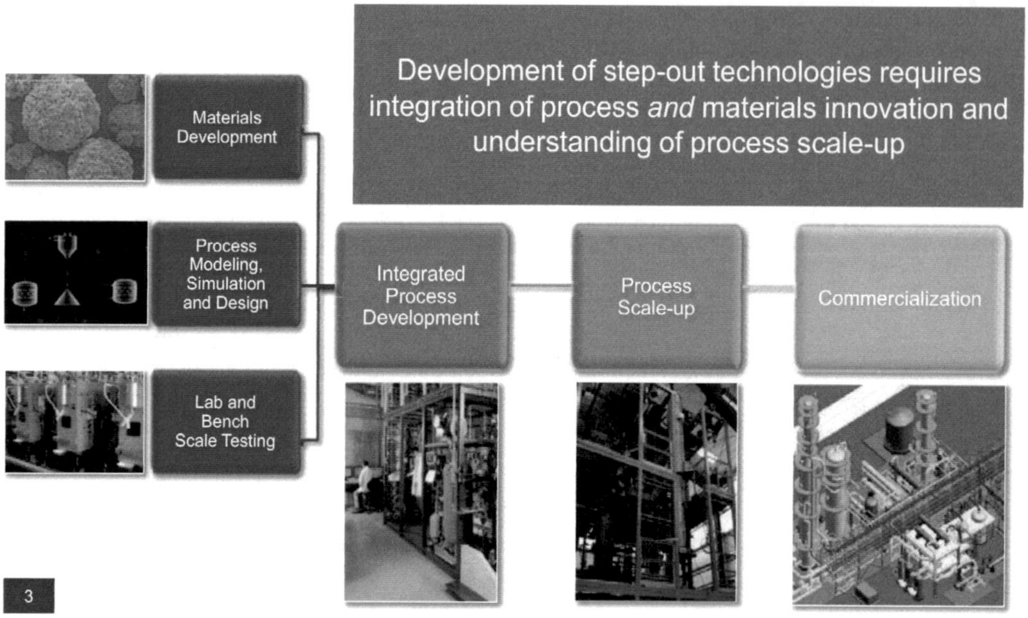

Application of Fluidized-bed Reactor Systems

- Rapid kinetics (residence time ~ 1-5 sec)
- Desired reactions are highly exothermic/endothermic
- Thermodynamic considerations
- Minimization of side reactions
- Ability to manufacture catalysts/adsorbents either as Group A or B particles
- Understanding of hydrodynamics in the reactor

Understanding of particle science is needed for successful process design. RTI INTERNATIONAL

Reactor Design Considerations

- Transport reactor vs. Bubbling-bed or a combination of two
- Single loop vs. double loop (once through or continuous regeneration)
- Cyclones and barrier filters
- Non-mechanical vs. mechanical valves
- Integration of strippers
- Y-legs vs. J-legs
- Gas distributors
- Aeration and dp measurements
- Temperature profiles
- Density profiles

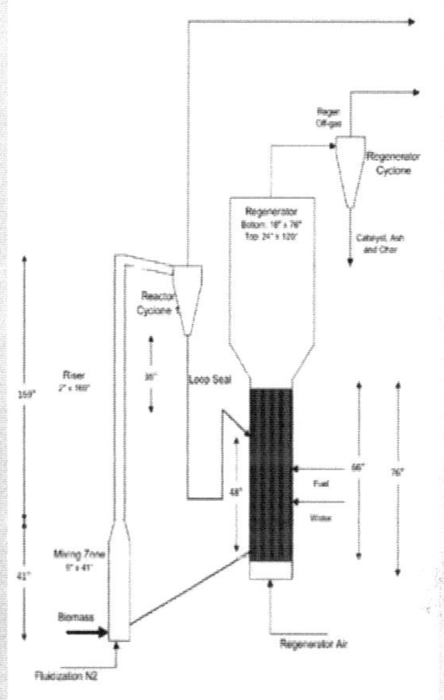

5

Particle Design Considerations

- Selection of active components and supports
- Dispersion of active/catalytic species on the support
- Particle size distribution
  - Group A: 40 to 120 μm
  - Group B: 100 to 300 μm
  - Concentration of fines
- Surface area
- Pore size distribution
- Attrition resistance
- Particle density
- Crystallographic phases (XRD)

Optimum combination of activity and mechanical strength is critical.

RTI INTERNATIONAL

6

Novel Technologies Developed by RTI

- Substitute natural gas (SNG) from syngas in a transport reactor
- Catalytic biomass pyrolysis to produce biofuels
- CO_2 capture from syngas and flue gas streams using solid adsorbents
- Desulfurization of coal-derived syngas in a dual-loop transport reactor

In all these technologies, careful design of catalyst/adsorbent particles is an enabler.

RTI INTERNATIONAL

Advanced SNG Process

- Coal to methane: 45% reduced capital cost compared to conventional fixed bed SNG

Coal to Substitute Natural Gas (SNG)

$CO + 3H_2 = CH_4 + H_2O$ $\Delta H = -206$ kJ/mol
$CO_2 + 4H_2 = CH_4 + 2\,H_2O$ $\Delta H = -165$ kJ/mol

To manage this ΔH, we need 4-6 stages of fixed-bed reactors with inter-stage cooling.

RTI INTERNATIONAL

RTI Advanced SNG Technology

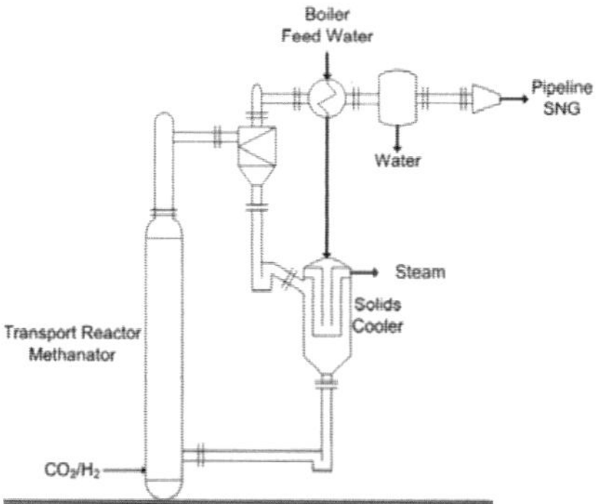

... based on a Fluidizable Methanation Catalyst

- High attrition resistance
- Methanation activity comparable to commercial fixed-bed methanation catalysts
- Catalyst production at pilot-scale with Clariant

RTI Transport Reactor SNG Process ...

- Targeted at coal to SNG conversion
- Exothermic temperature control and effective heat recovery
- Small footprint for high throughput

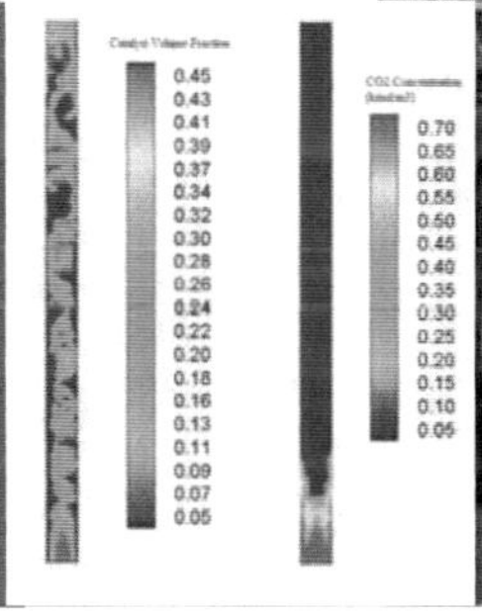

RTI Advanced SNG Technology

Transport Reactor Methanation Process

- Detailed techno-economic assessment with leading commercial partner in methanation technology
  - Focus on methanation block
  - Comparison of conventional fixed bed with RTI transport reactor design
- ~ 45% lower CAPEX than conventional fixed-bed systems
- Syngas conversion cost to SNG reduced by ~ 20%

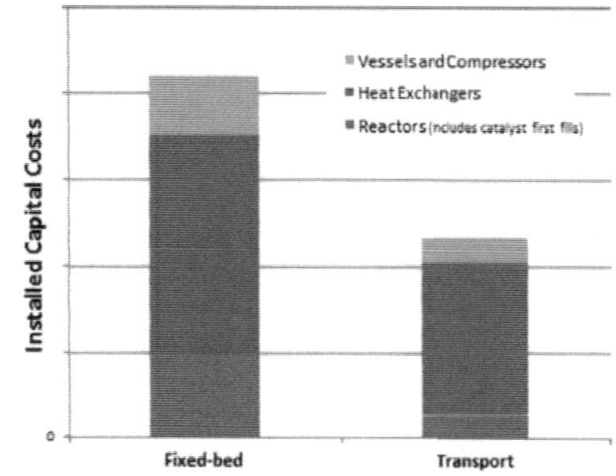

RTI INTERNATIONAL

RTI International | Energy Technology Division

Biomass Pyrolysis

- Producing stable, economically viable bio-crude intermediate suitable for conventional refining infrastructure

11

RTI International | Energy Technology Division

Catalytic Biomass Pyrolysis

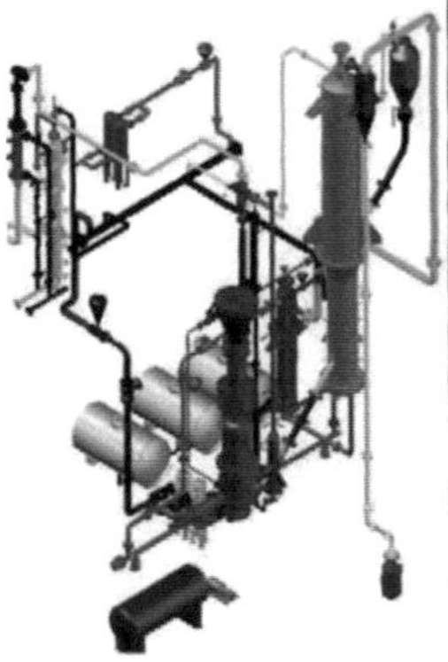

RTI Transport Reactor Catalytic Biomass Pyrolysis Process...

- Produces "bio-crude" intermediate from lignocellulosic biomass that can be processed in existing refineries
- Single loop transport reactor system, continuous catalyst reaction and regeneration
- System can be operated auto-thermally with heat of regeneration

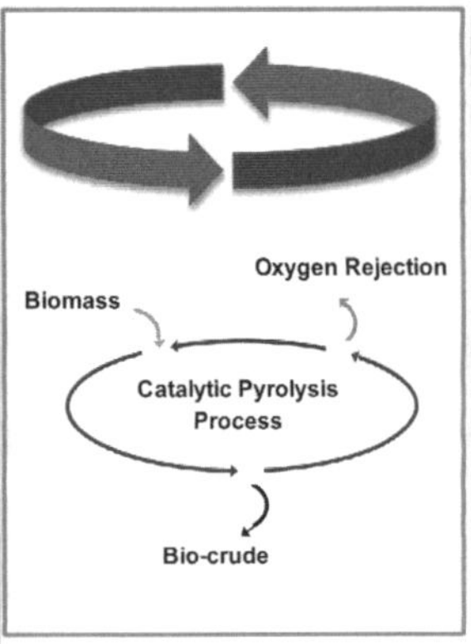

... based on a Multi-functional Fluidizable Catalyst

- Multi-functional catalyst to maximize carbon efficiency, remove oxygen, and control bio-crude properties
- High attrition resistance

RTI

12

Why transport reactor for catalytic biomass pyrolysis?

- Short residence time of pyrolysis vapors in reactor
- Longer residence time of catalyst in regeneration
- Continuous catalyst regeneration to overcome rapid deactivation by carbon deposition in the process
- No need to interrupt solids feed to system for catalyst regeneration
- Maintain narrow temperature window to maximize liquid product yield
- Improved heat integration allowing heat generated in catalyst regeneration to be transferred to maintain temperature in the pyrolysis reactor.

RTI 1 ton per day catalytic biomass pyrolysis pilot unit.

RTI INTERNATIONAL

13

Summary of Technology Status

- Catalyst development included model compound screening and bench-top (~ 1 g/hr) biomass conversion
- Catalytic biomass pyrolysis in a 1"-dia fluidized bed reactor
  - 20 wt% oxygen content with 42% energy recovery
- Laboratory data provides the basis for a 1 TPD bench-scale unit
- 1 TPD bench-scale unit operational
- Preliminary Techno-Economic Analysis
- Preliminary Bio-crude Upgrading
- Next Steps: Extended Pyrolysis and Upgrading
- Technology commercialization

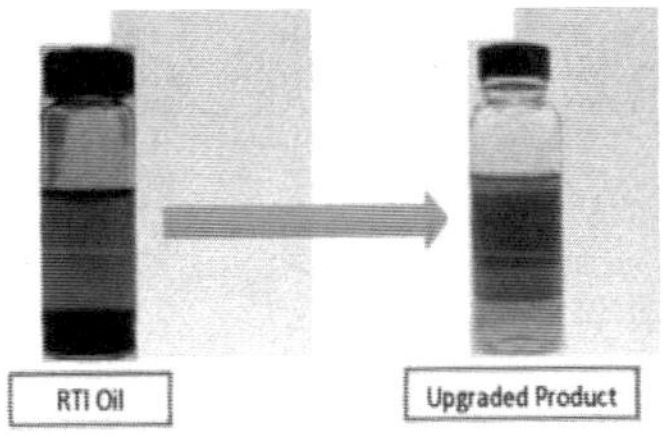

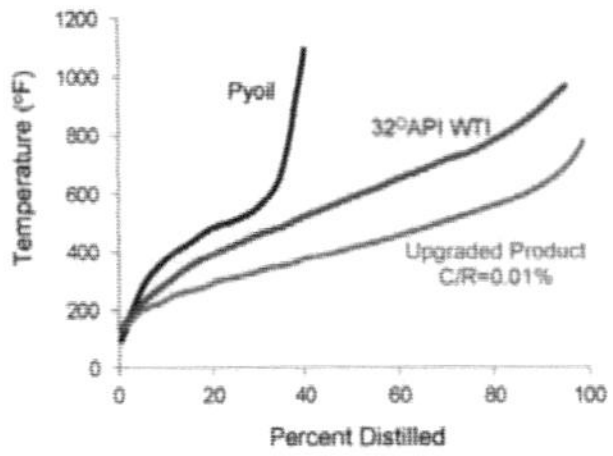

14

RTI International

Advanced Sorbent CO_2 Capture Process

RTI is developing a sorbent-based CO_2 capture process that uses a supported, polymeric amine sorbent and a fluidized-bed process arrangement for post-combustion applications (power and industrial installations).

- Technology in development at RTI and PSU since ~2000
- Strong partnerships with government and industrial entities

NETL, United States Environmental Protection Agency, PennState, CLARIANT, Masdar

Current development efforts:

- transition a promising sorbent chemistry to a low-cost sorbent suitable for use in circulating fluidized-bed process
- bench-scale demonstration of complete circulating fluidized-bed process

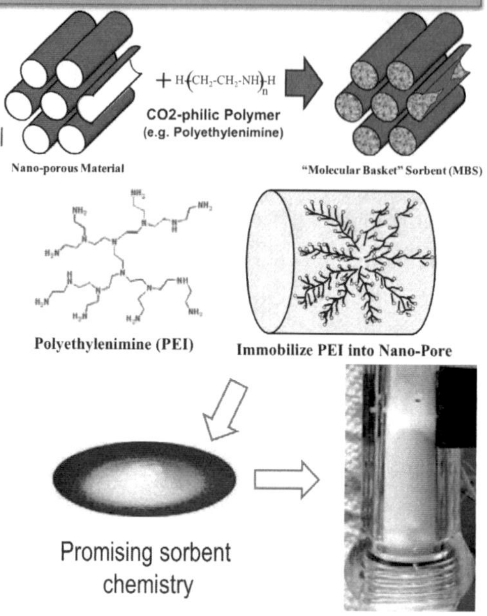

15

RTI International — Energy Technology Division

Conversion to a Fluidizable Platform

Converted powder sorbents to low-cost, fluidizable, attrition-resistant particle suitable for use in a fluidized-bed process

- Spray Drying
- Suitable particle size and density
- Physically strong and resistant to attrition
- Non-cohesive

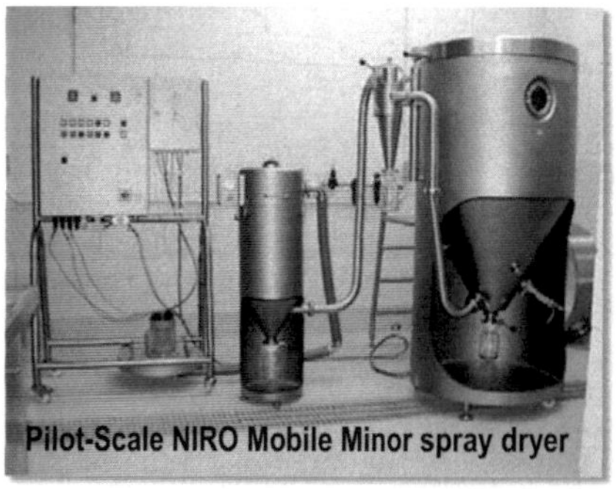
Pilot-Scale NIRO Mobile Minor spray dryer

Evaluation of fluidizability under process conditions

- Lab-scale, visual fluidized bed constructed to evaluate prepared sorbents at specific process conditions to allow for visual observation / characterization of the sorbent bed

RTI International

16

Fluidized-bed Process Arrangement

Conceptual Process Arrangement:

Circulating, Staged, Fluidized-bed Reactor with Internal Heat Management

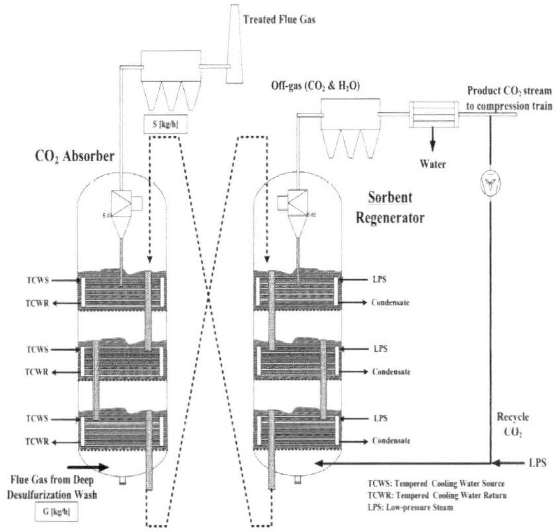

Benefits

- Mimics conventional gas-liquid absorption processes
- Counter current gas-solids flow maximizes CO_2 driving force throughout reactor length
- Bed staging effectively enables counter-current flow
- Superior gas-solid heat and mass transfer characteristics and heat management strategy minimize thermal regeneration energy
- Reduced pressure drop in fluidized state

Development Needs

- Optimize reactor design and process arrangement

Development Approach

- Detailed fluidized bed reactor modeling
- Bench-scale evaluation of reactors designs
- Demonstration of process concept

RTI INTERNATIONAL

17

Fluidized-bed Reactor Modeling

Approach

- Based on extensively validated 1-dimensional, semi-empirical, two-phase fluidized-bed reactor model[1]
- Model was selected since it had been developed to design and scale-up industrial fluidized-bed processes
- Modifications include
  - continuous addition and removal of sorbent to fluidized bed
  - accumulation term to account for the CO_2 buildup or release
  - heat transfer between the fluid bed and internals
- Model Input
  - sorbent particle properties (particle size, density, sphericity, Geldart type, heat capacity, thermal conductivity)
  - CO_2 capture kinetics and CO_2 isotherms (equilibrium)
  - reactor geometry (dimensions, staging, heat exchangers, etc.)
  - operating conditions (temperature, pressure, feed rates, etc.)

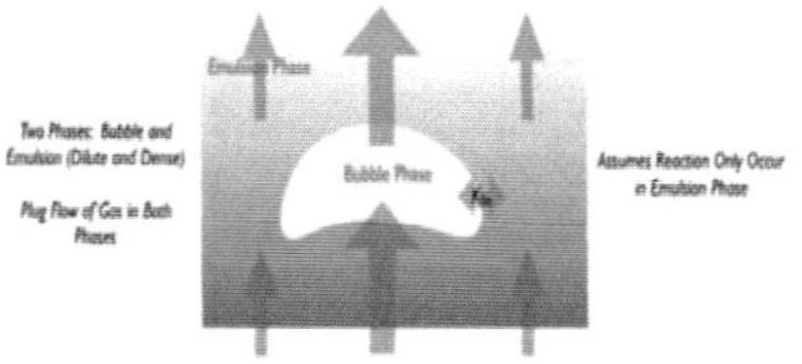

What will the model help us do?

- Understand the effect of key process and sorbent parameters on the performance of the proposed FMBR designs
- Optimize design of CO_2 Absorber and Sorbent Regenerator including heat transfer internals and bed-staging

FB Reactor Development Plan

Experimental eval. & char. of bench-scale FB reactors → Collection of experimental data to validate FBR model → FBR model tuning and prediction of optimal operating conditions → Experimental demonstration / verification of 'optimal' reactor operation

[1]Modeling of Fluidized Bed Reactors; Joachim Werther, Ernst-Ulrich Hartge; International Journal of Chemical Reactor Engineering, 1 (2003) P1

RTI INTERNATIONAL

18

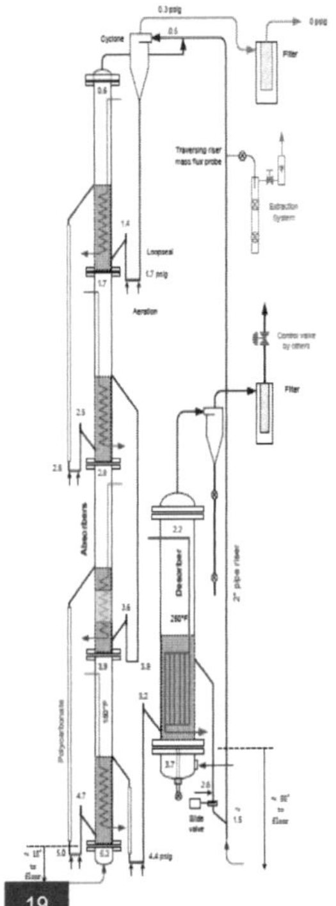

Bench-scale Contactor Evaluation Unit

Staged, Circulating Fluidized-bed with Integrated Heat Management

- Adequately sized to avoid issues related to bed slugging that would lead to poor CO2 capture performance and possible damage to the columns
- evaluate effectiveness of staged reactor design with integrated heat management system for CO_2 removal from flue gas
- evaluate fluidizable, attrition-resistant , supported polymeric amine sorbent

Specifications

- Flue gas throughput: 300 and 900 SLPM
- Solids circulation rate: 75 to 450 kg/h
- Sorbent inventory: ~100 kg of sorbent

- **Currently being commissioned**

RTI INTERNATIONAL

19

Warm-Syngas Clean-Up Technology

- 20+% percent reduction in capital cost compared to conventional low temperature processes
- Improved thermal efficiency (up to 3% HHV)

20

Gasification is Expensive: How to Reduce Its Cost?

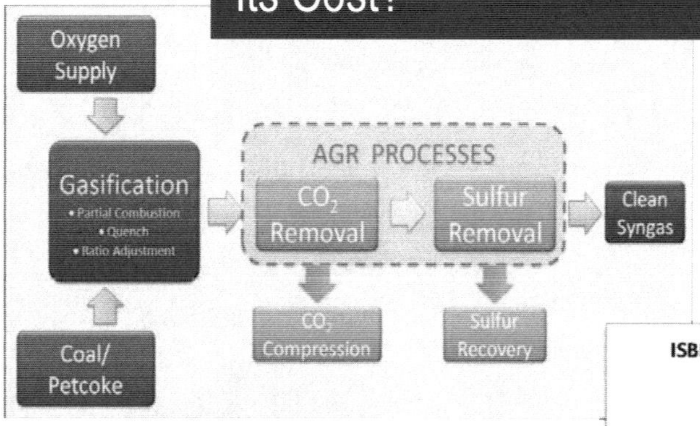

2. Make Efficiency Improvements that Reduce all Elements of Cost

1. Reduce Cost of Major Process Blocks, such as Syngas Cleanup

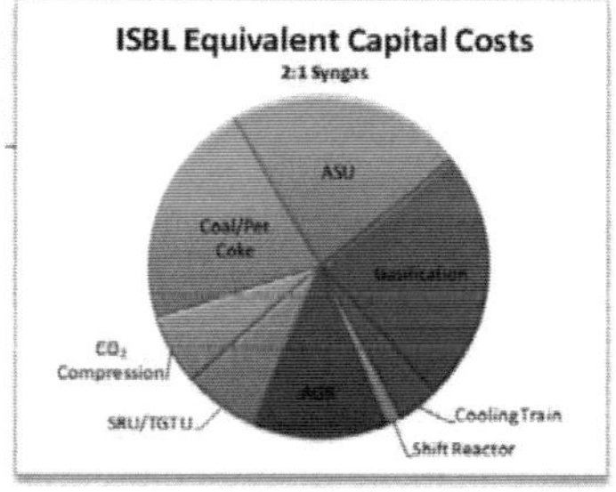

RTI's Focus is to Both Improve Efficiency and Reduce Costs for Optimum Benefit!

Graphics Source: Eastman Chemical Company

RTI INTERNATIONAL

21

Warm Syngas Desulfurization Process

A unique process technology based on transport reactor design (similar to a commercial FCC reactor design]...

... based on the development of a highly active, attrition resistant sorbent.

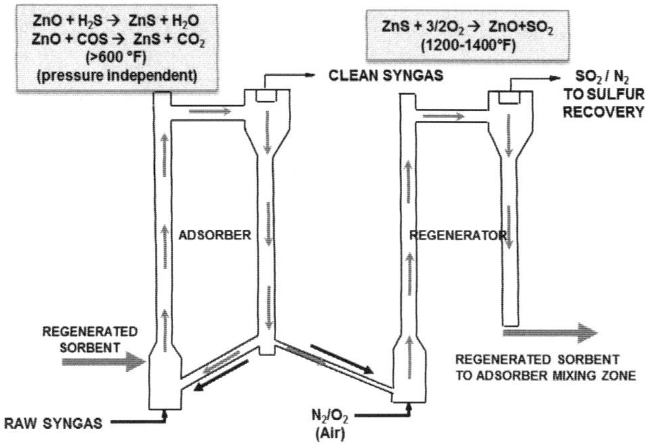

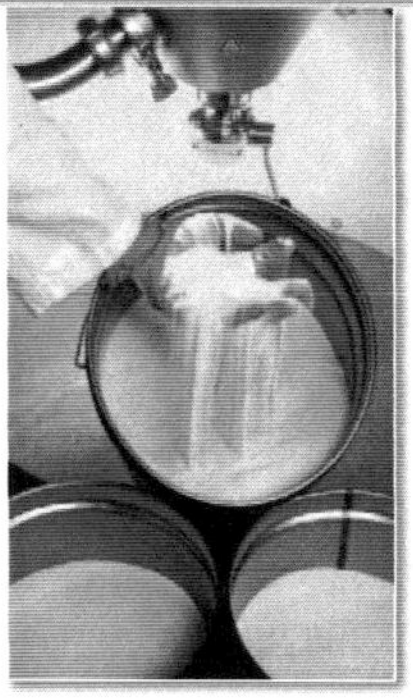

RTI proprietary desulfurization sorbent

- R&D 100 Award in 2004
- Developed through long-term cooperation with Clariant

Part of comprehensive high temperature contaminant removal platform.

RTI INTERNATIONAL

22

Pilot-Scale Field Testing at Eastman

- Warm Desulfurization Process (WDP)
  - > 99.9% removal of H_2S and COS (< 5 ppmv effluent S) at temperatures > 600°F and pressures up to 600 psig
  - > 3,000 hours of operation and parametric testing
  - Lower attrition than commercial FCC systems
  - Commercial-scale sorbent production at Süd Chemie (now Clariant)
- Direct Sulfur Recovery Process (DSRP)
  - > 99.8% SO_2 conversion to elemental sulfur
- Trace Contaminant Removal Process (TCRP)
  - > 95% NH_3/HCN removal with regenerable sorbent
  - > 90% Hg and As removal with fixed-bed sorbents

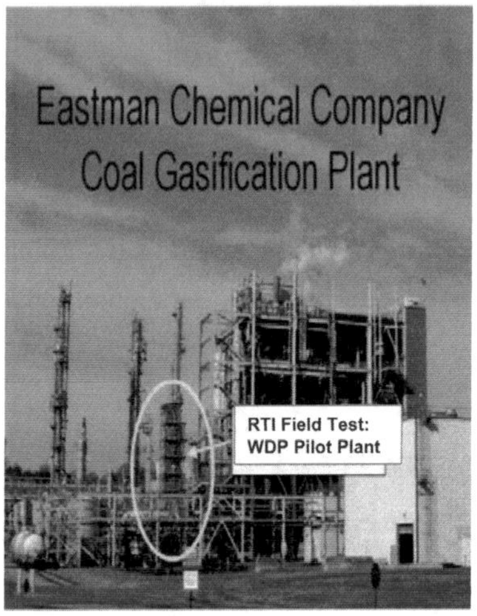

Picture Source: Eastman Chemical Company

RTI INTERNATIONAL

WDP Scale-Up Factors

	EMN Pilot Plant	Demonstration	Commercial
Size (MWe)	**0.3**	**50***	**600****
Gas Flow (KSCFH)	8	2,000	24,000
Footprint (ft^2)	260	1,100	3,000
Adsorber Mixing Zone Riser	 15'H x 2.5"ID 40'H x 1.5"ID	 16'H x 54"ID 50'H x 28"ID	 16'H x 15'ID 50'H x 8'ID
Regenerator Mixing Zone Riser	 10'H x 1.5"ID 20'H x 1"ID	 21'H x 14"ID 44'H x 8"ID	 21'H x 4'ID 44'H x 2.5'ID

* About half the size of commercial industrial gasification systems which are typically ~100 MWe.

** Approximately the size of current state-of-the-art IGCC reference plants (~600 MWe)

RTI INTERNATIONAL

Comparison of Attrition Performance for RTI-3 vs. FCC Catalyst

PSRI

PSRI Jet Cup Test – Threshold Velocity

Grace FCC Equilibrium Catalyst (Genesis®)

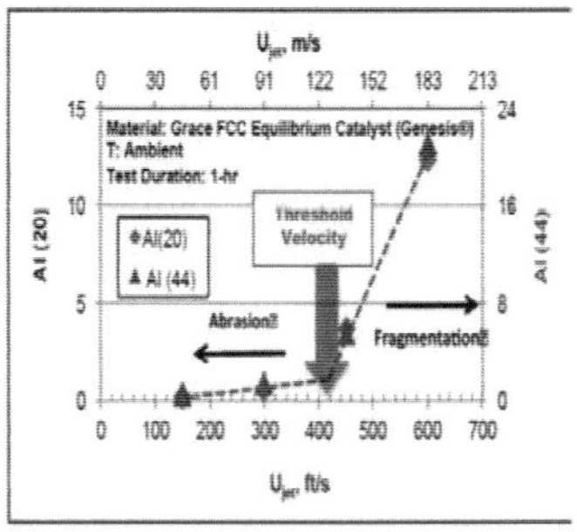

- Threshold velocity - the velocity at which the slope of the attrition index vs. velocity line changes
- Threshold velocity was somewhere around 410 ft/s (125 m/s) for Grace FCC equilibrium catalyst (Genesis®) at ambient conditions

PSRI

PSRI Jet Cup Test – Threshold Velocity

RTI Sorbent Catalyst (Clariant Al_2O_3+ZnO)

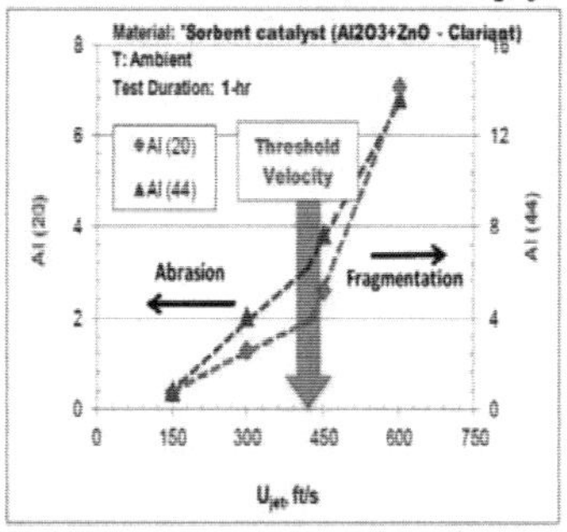

- Threshold velocity - the velocity at which the slope of the attrition index vs. velocity line changes
- Threshold velocity was somewhere around 420 ft/s (128 m/s) for Sorbent Catalyst at ambient conditions

Proprietary and Confidential

RTI INTERNATIONAL

26

Process scale-up steps

Process Design

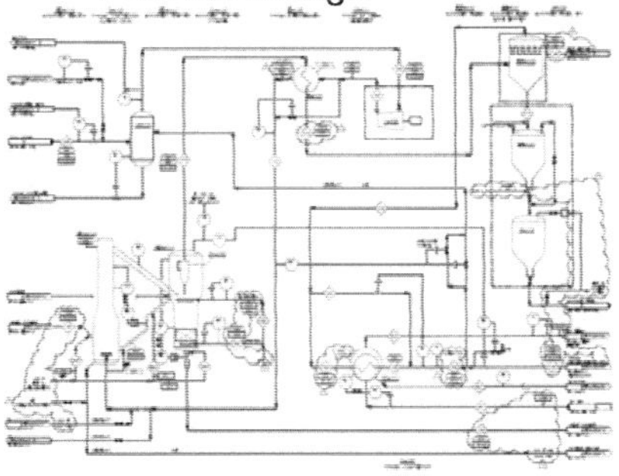

CFD Modeling

Instantaneous Particle and Temperature

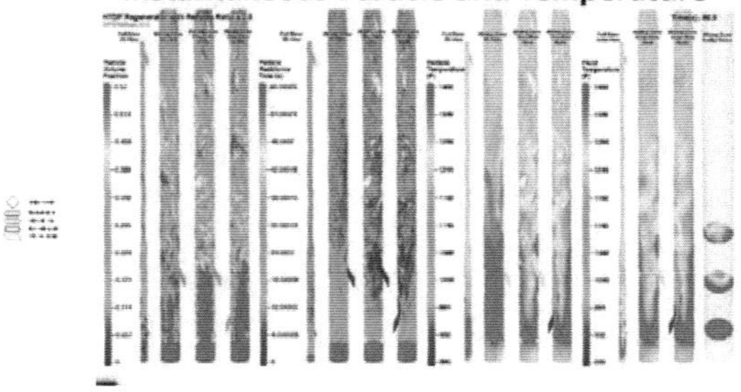

Reactor Internals

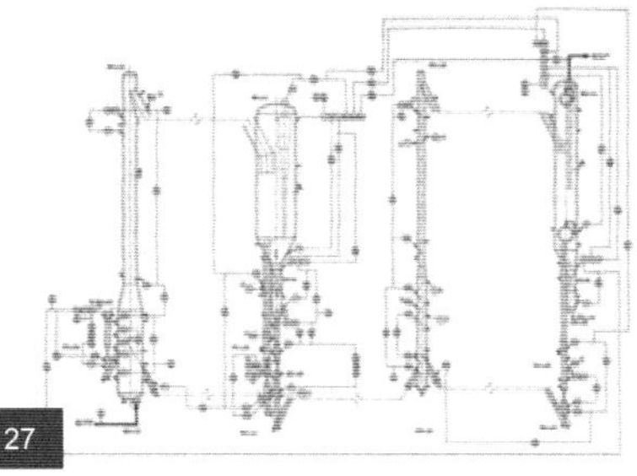

Long-term Sorbent Testing

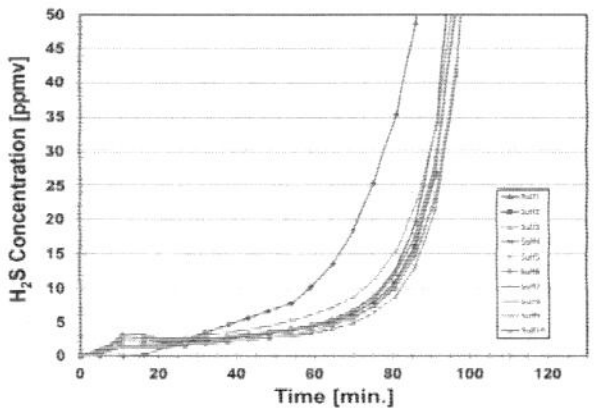

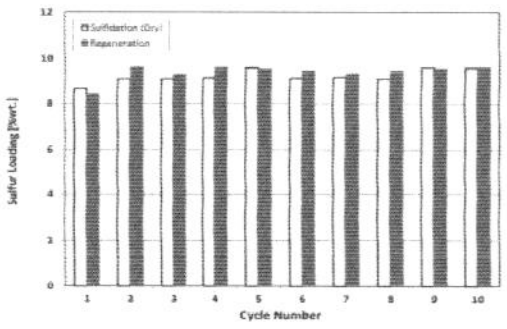

RTI INTERNATIONAL

27

RTI Warm Syngas Desulfurization Technology – Scale-up History

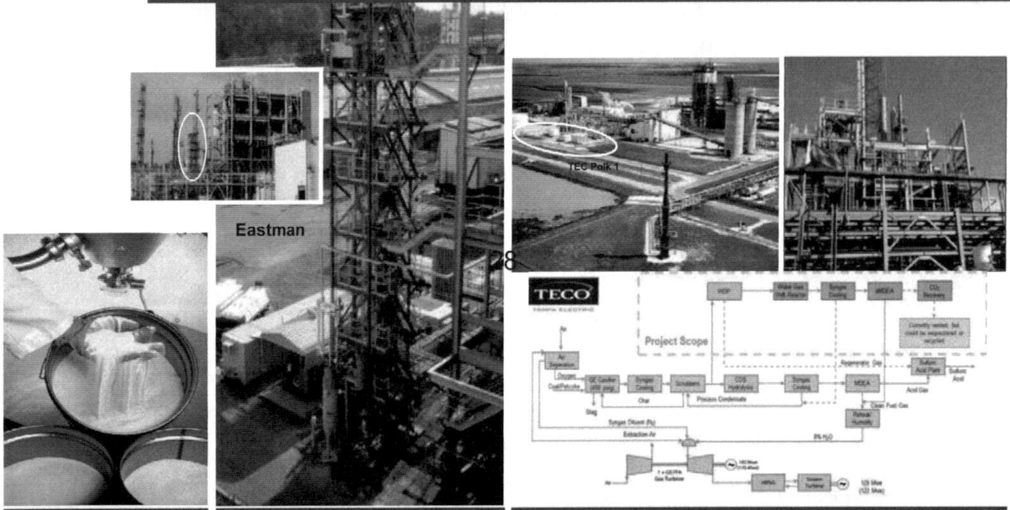

Invention (2001)
- Proprietary Desulfurization Sorbent - R&D 100 Award (2004)

Pilot testing (2006-2008)
- Eastman Chemical Company
- 3000hr pilot test, coal-derived syngas
- > 99.9% removal of H_2S and COS (< 5 ppmv S) at > 600°F & > 600 psig

Demonstration (2010-2015):
- Tampa Electric IGCC Plant, Florida
- $168.8MM DOE funding to design, construct, operate
- 50 MW_e equivalent scale (~20% syngas slip-stream)

28

RTI International Energy Technology Division

RTI 50-MWe Warm Syngas Cleanup Demonstration Project -Construction

Installing Warm Syngas Desulfurization Equipment

Installing WGS Reactors

Completed Construction – 50-MWe Demonstration Project

Installing Carbon Capture System

Opening Ceremony on April 9, 2014

29

Summary

- Fluidized-bed reactors play a critical role in clean energy technologies.
- Successful development requires concurrent development of particles and reactor systems.
- Careful coupling of hydrodynamics and kinetics is a key to scaling up the reactor systems.
- Operation at larger scale is easier than at small lab or bench-scales.
- Higher pressures generally result in better quality of fluidization.

Cheap natural gas (in place of oil) can be used to generate higher value products – the price spread with oil supports job creation

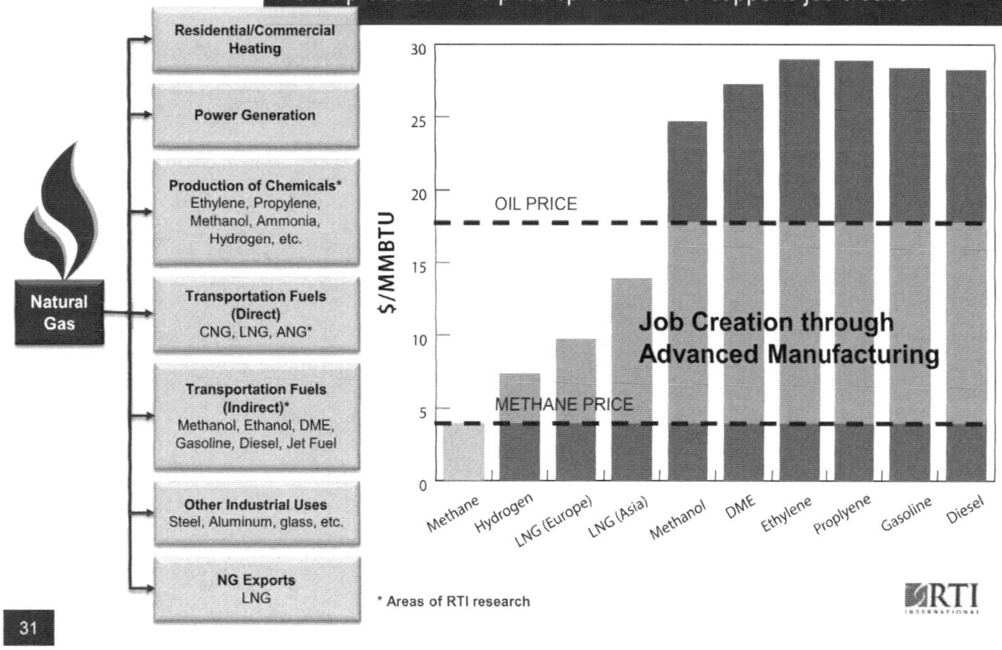

* Areas of RTI research

RTI International
Our team...
Raghubir Gupta, Vice President
Energy Technology Division
RTI International
gupta@rti.org
+1-919-541-8023
32

Lab and In-line 3-Dimensional Image Analysis of Particulate Powders and Slurries.

A MEMBER OF NIKKISO

Microtrac
Total Solutions in Particle Characterization

What Is Image Analysis?

A MEMBER OF NIKKISO

Microtrac
Total Solutions in Particle Characterization

What Is Image Analysis?

- ✓ *Photos* of individual particles are digitized, stored in a viewable image file
- ✓ Based on the size of a pixel, dimensions are measured
- ✓ Further calculations are performed to characterize better.
- ✓ Static IA – Microscopy
- ✓ Dynamic IA – Particle in motion

A MEMBER OF NIKKISO

Microtrac
Total Solutions in Particle Characterization

Why Image Analysis?

✓One picture is worth a thousand words...

✓Provides more information to characterize individual particles.

✓Adds to the information provided by diffraction technology

4

A MEMBER OF NIKKISO

Microtrac
Total Solutions in Particle Characterization

When Do I Use It?

✓Evaluate distribution change

✓Evaluate sample preparation- Qualitative evaluation replaced with quantitative measurement

✓R&D – Part of an overall particle characterization.

✓Replace microscopy – a difficult quantitative analysis replaced with an easy quantitative analysis

Dry System - How It Works

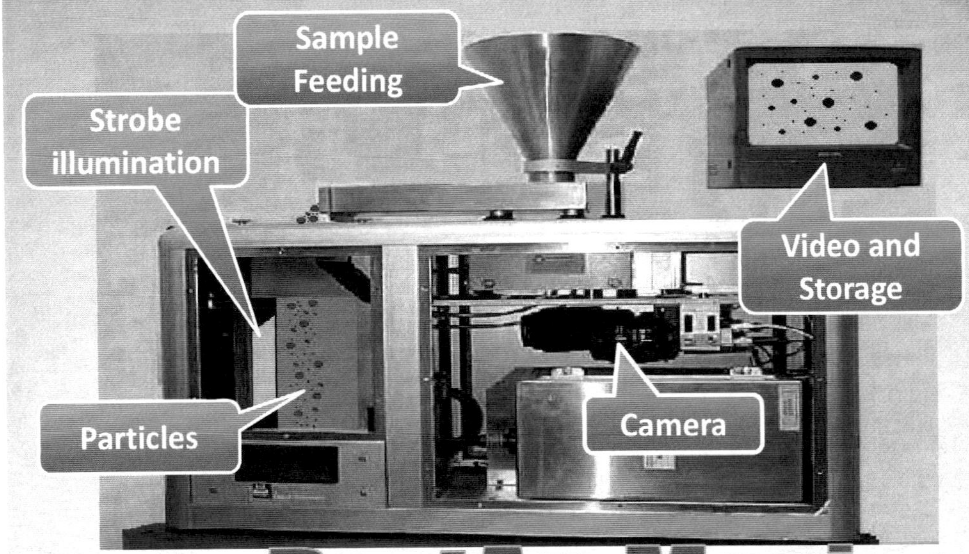

PartAn Maxi

Images
Line around particle indicates that it is included in the measurement
A MEMBER OF NIKKISO
Microtrac
Total Solutions in Particle Characterization

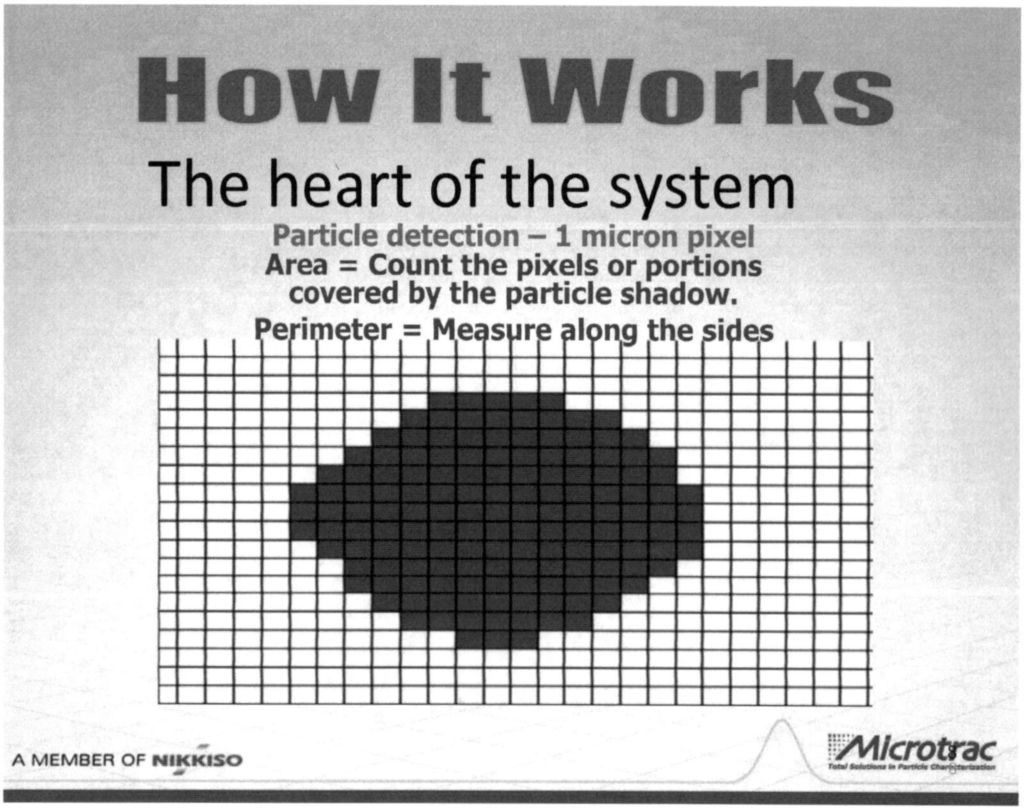
How It Works
The heart of the system
Particle detection – 1 micron pixel
Area = Count the pixels or portions covered by the particle shadow.
Perimeter = Measure along the sides
A MEMBER OF NIKKISO
Microtrac
Total Solutions in Particle Characterization

Calculations

- ✓ Various geometric parameters (area, length, width, perimeter)
- ✓ Calculate shape factors (ratios of sizes like width/length = aspect ratio)
- ✓ Report data as graphs, tables, summary data of all parameters

A MEMBER OF NIKKISO

Microtrac
Total Solutions in Particle Characterization

PartAn Maxi - dry powder analysis using 3D imaging

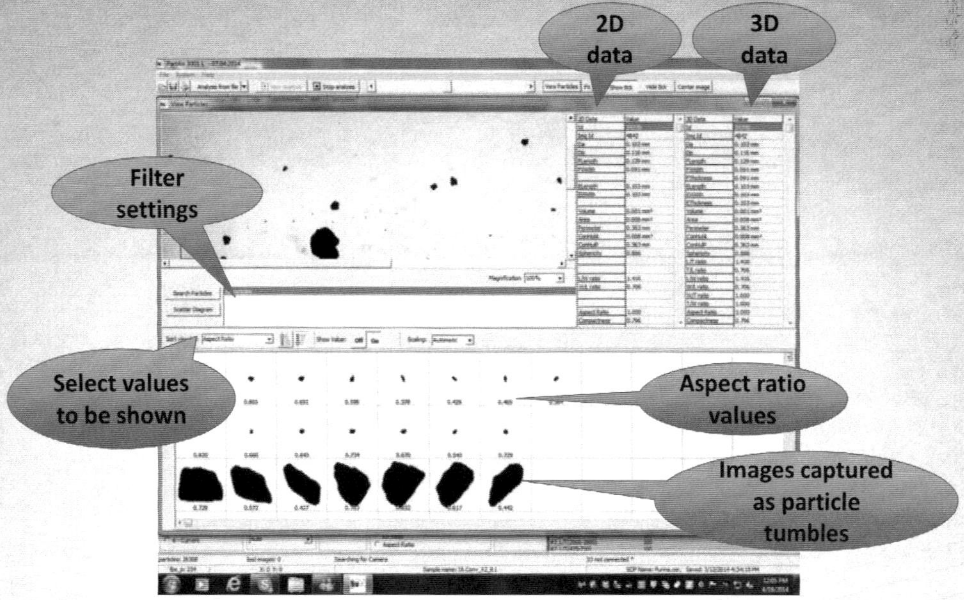

A MEMBER OF NIKKISO

Microtrac
Total Solutions in Particle Characterization

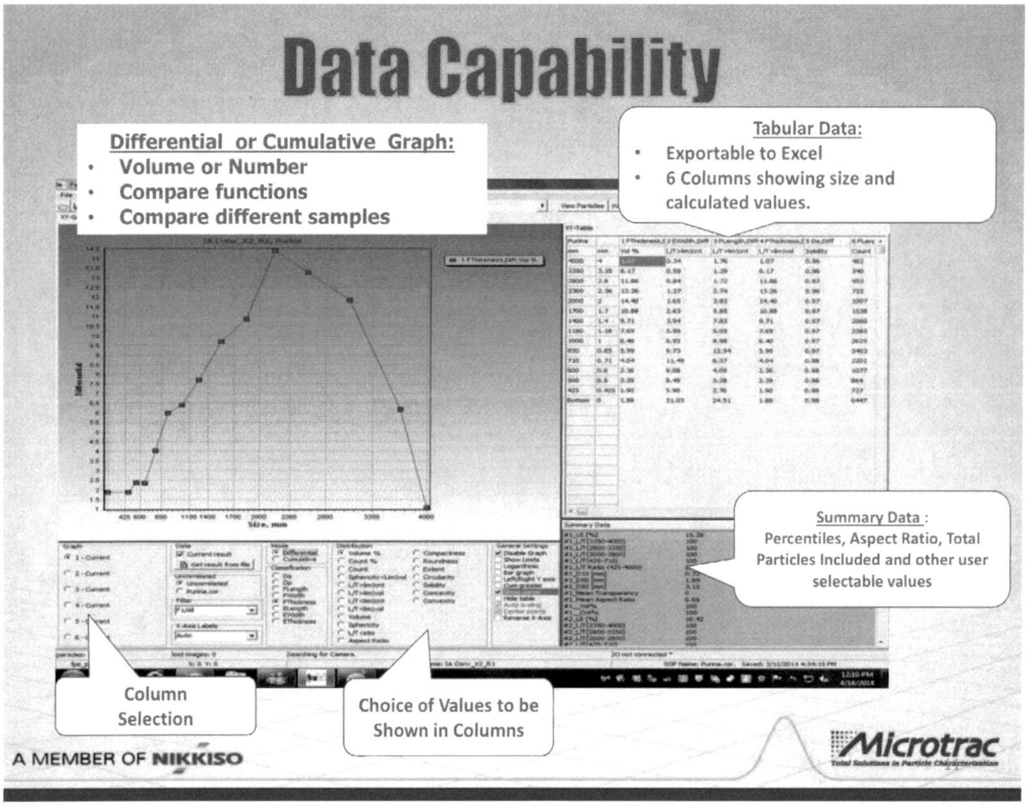
Data Capability
Differential or Cumulative Graph:
• Volume or Number
• Compare functions
• Compare different samples
Tabular Data:
• Exportable to Excel
• 6 Columns showing size and calculated values.
Summary Data : Percentiles, Aspect Ratio, Total Particles Included and other user selectable values
Column Selection
Choice of Values to be Shown in Columns
A MEMBER OF NIKKISO
Microtrac
Total Solutions in Particle Characterization

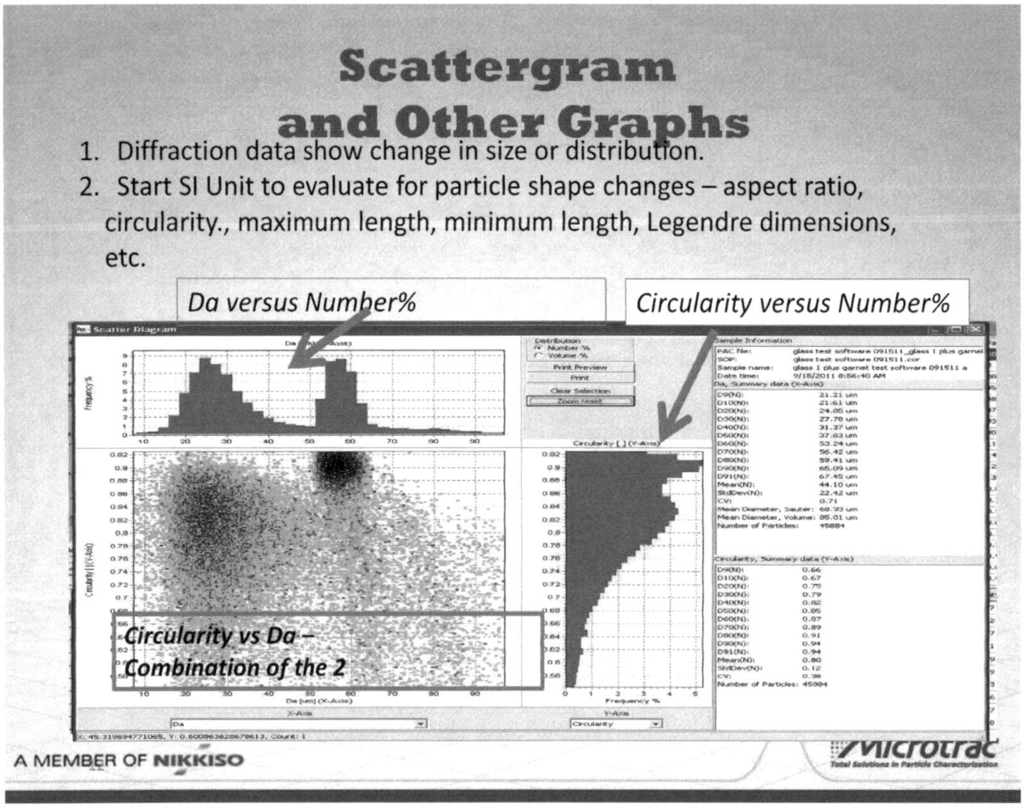
Scattergram and Other Graphs
1. Diffraction data show change in size or distribution.
2. Start SI Unit to evaluate for particle shape changes – aspect ratio, circularity., maximum length, minimum length, Legendre dimensions, etc.
Da versus Number%
Circularity versus Number%
Circularity vs Da – Combination of the 2
A MEMBER OF NIKKISO
Microtrac
Total Solutions in Particle Characterization

Data Capability

- ✓ 14 Size and 12 Shape parameters
- ✓ Tabular and Graphical Volume and Number (count) frequency distributions of all parameters as differential or cumulative plots
- ✓ Summary data for all parameters – mean, percentiles, Std Dev, etc
- ✓ Scattergram graph for all images including ability to filter, and create morphological classifications with, all parameters
- ✓ User-assignable standard operating procedures
- ✓ Automatic Report printouts are user selectable
- ✓ Ability to export all data in spreadsheet form
- ✓ Correlation capability to other sizing techniques – e.g. sieves
- ✓ Graph overlay capability

A MEMBER OF NIKKISO

Microtrac
Total Solutions in Particle Characterization

3-Dimensional Measurements-

Example: Measurement of a coin

1. Image used for thickness

2. Image used for Da, Length etc.

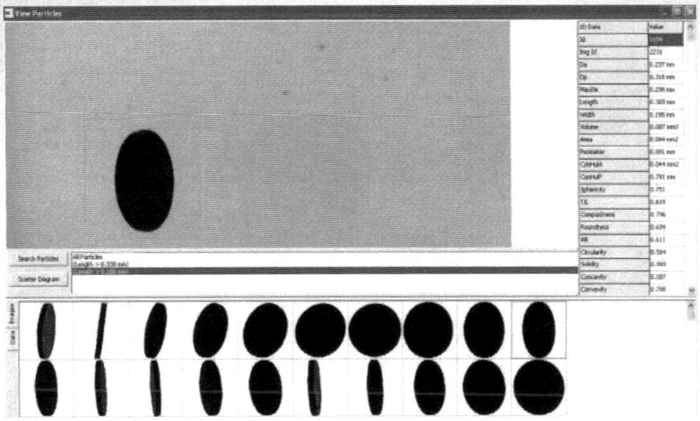

A MEMBER OF NIKKISO

Microtrac
Total Solutions in Particle Characterization

SI Features - Fluids -

◆Any application that can be done by Microtrac diffraction: any fluid! – any particle!

◆Particle size and shape analysis 5 to 2000 Micron

◆Provides images of sample particles being measured

◆Often same concentration and sample for both

◆Operate separately or sequentially

◆Can be added to existing S3000, S3500 and Bluewave

◆Complete shape analysis and distribution

How SI Works

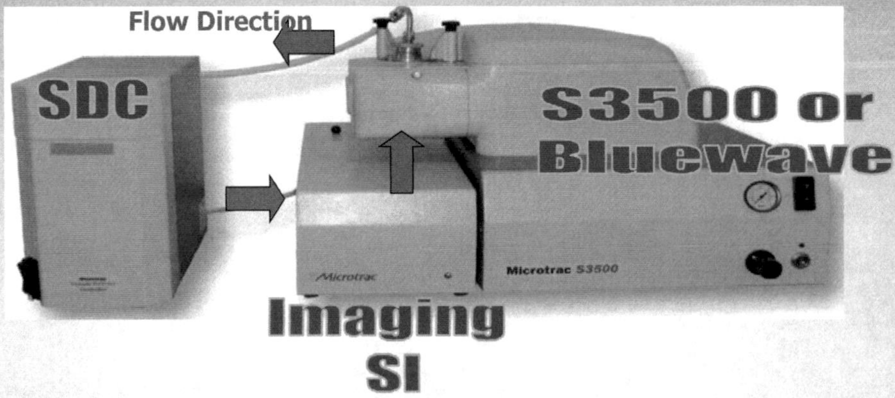

SI modular design allows connection to any S3000, S3500 or Bluewave

Measurement Principle Microtrac DIA / DIAF and SI

- ✓ White Light (strobe effect) illuminates particles
- ✓ Feeding system provides particles to light beam
- ✓ Optics (Lens) focuses image on CCD
- ✓ CCD Camera captures and digitizes images
- ✓ Image Measurement (we know how big a pixel is)
- ✓ Save the Image to file

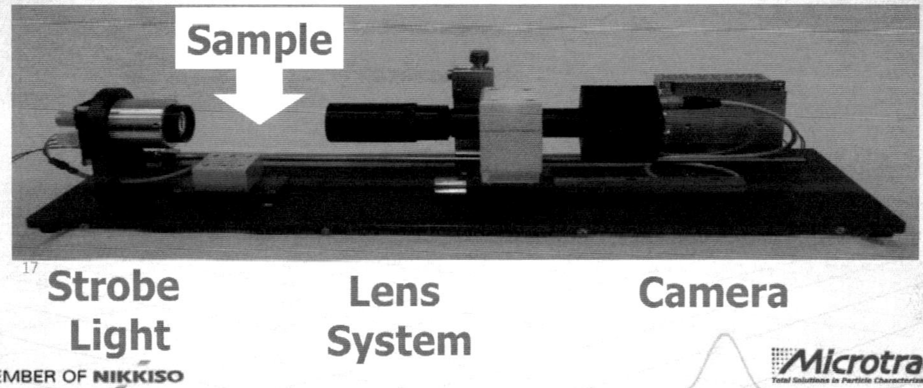

17

A MEMBER OF NIKKISO

Microtrac
Total Solutions in Particle Characterization

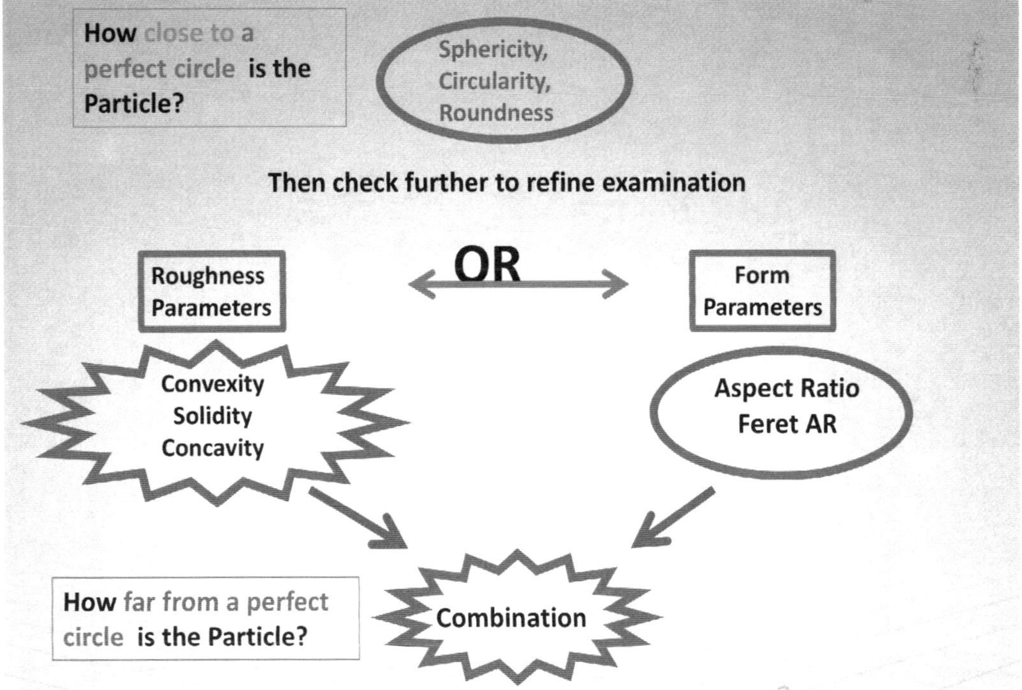

A MEMBER OF NIKKISO

18

Microtrac
Total Solutions in Particle Characterization

Imaging and Diffraction Data May not Agree !

THE MEASURING TECHNOLOGIES ARE DIFFERENT

Diffraction uses a <u>light scattering pattern</u> and performs calculations to convert it to particle size.

Imaging uses <u>pictures of particles</u> to measure basic dimensions and performs calculations using those dimensions.

A MEMBER OF NIKKISO

Microtrac
Total Solutions in Particle Characterization

ON_LINE Dry Powder Measurements

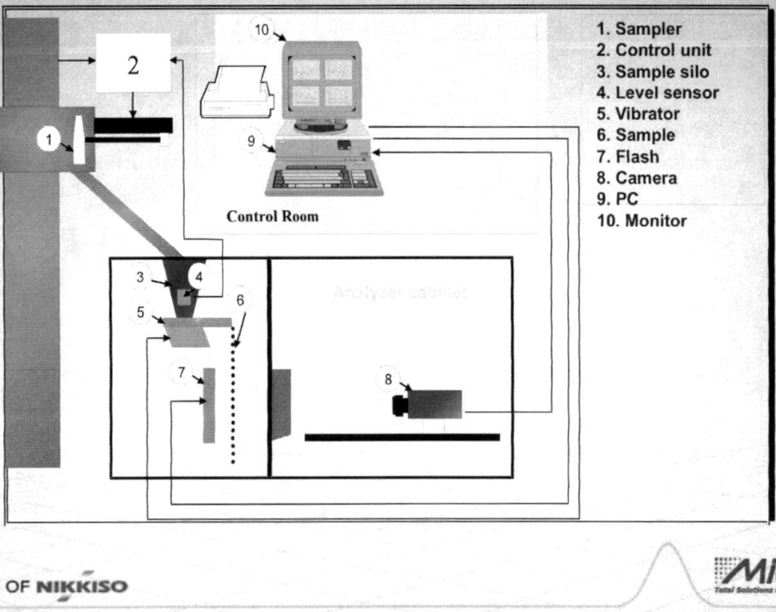

A MEMBER OF NIKKISO

Microtrac
Total Solutions in Particle Characterization

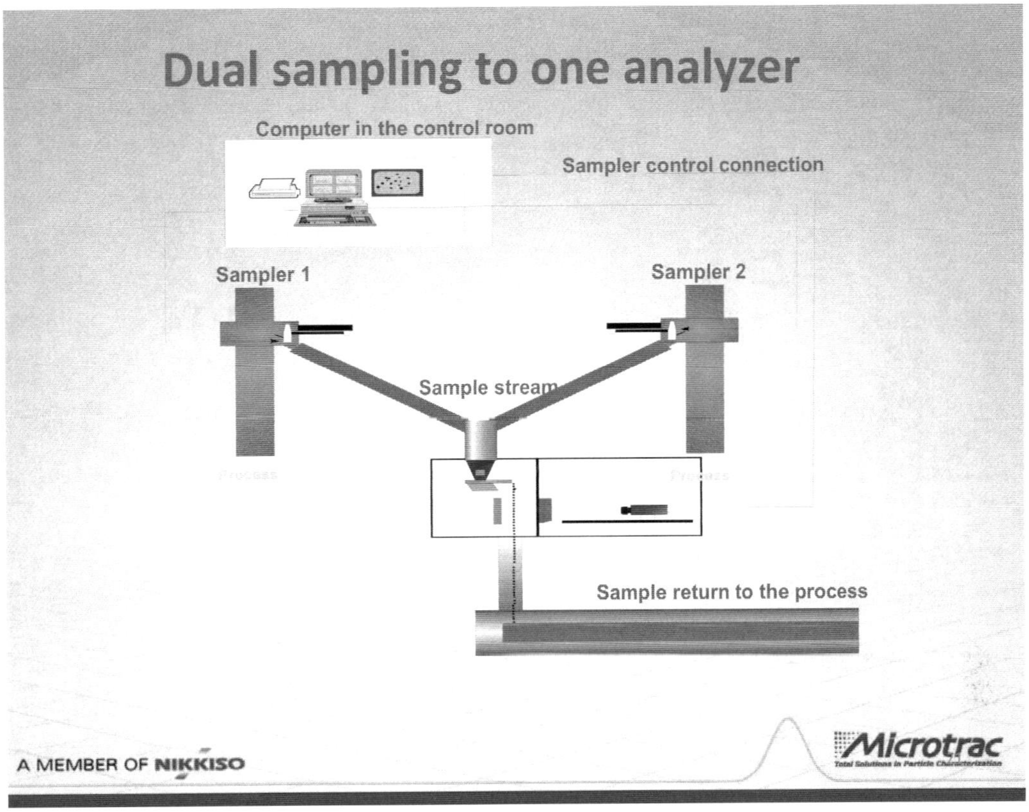
Dual sampling to one analyzer
Computer in the control room
Sampler control connection
Sampler 1
Sampler 2
Sample stream
Sample return to the process
A MEMBER OF NIKKISO
Microtrac

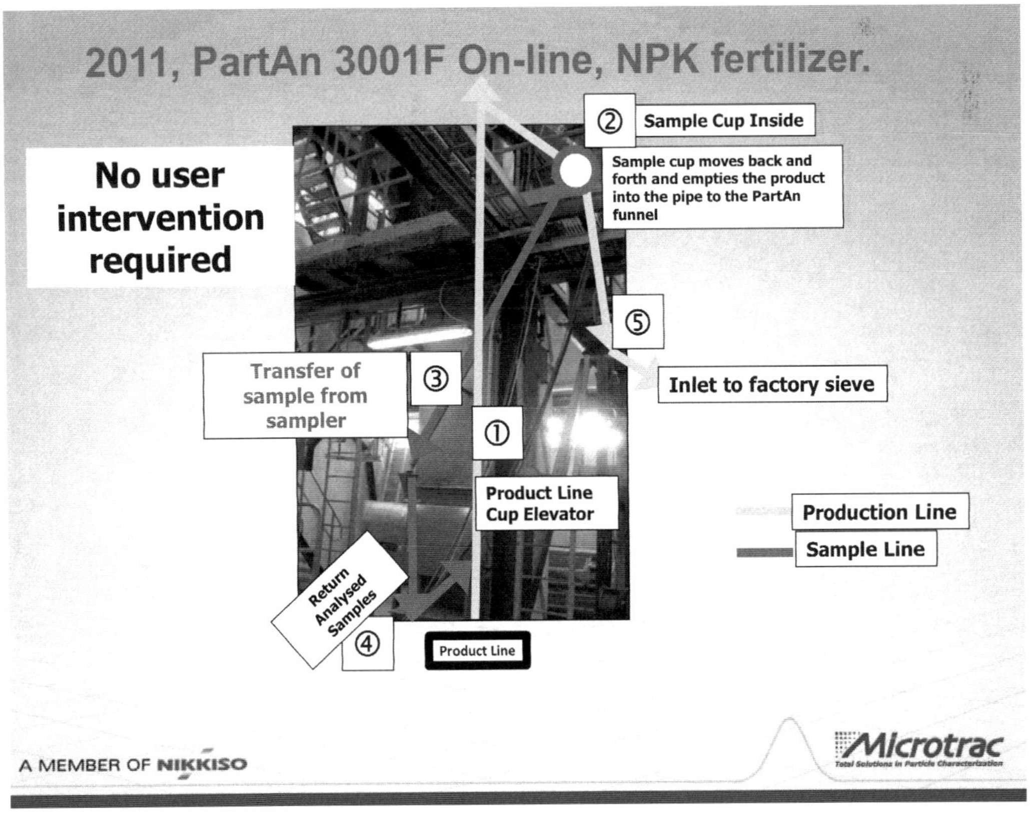
2011, PartAn 3001F On-line, NPK fertilizer.
No user intervention required
② Sample Cup Inside
Sample cup moves back and forth and empties the product into the pipe to the PartAn funnel
⑤
Inlet to factory sieve
Transfer of sample from sampler
③
①
Product Line Cup Elevator
Production Line
Sample Line
Return Analysed Samples
④
Product Line
A MEMBER OF NIKKISO
Microtrac

Suspension Measurements

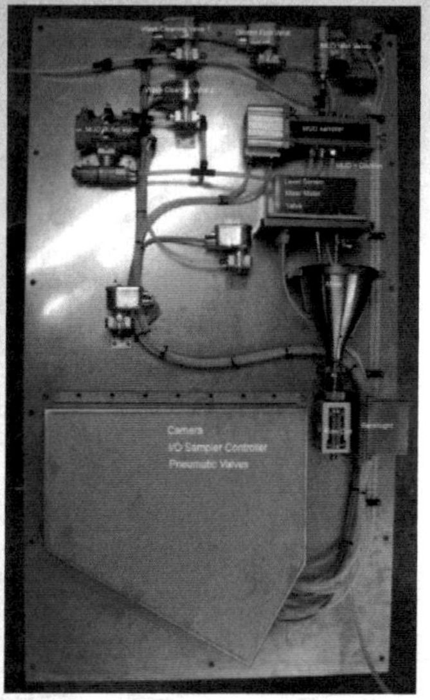

A MEMBER OF NIKKISO

Microtrac
Total Solutions in Particle Characterization

Suspension Measurements – another approach

A MEMBER OF NIKKISO

Microtrac
Total Solutions in Particle Characterization

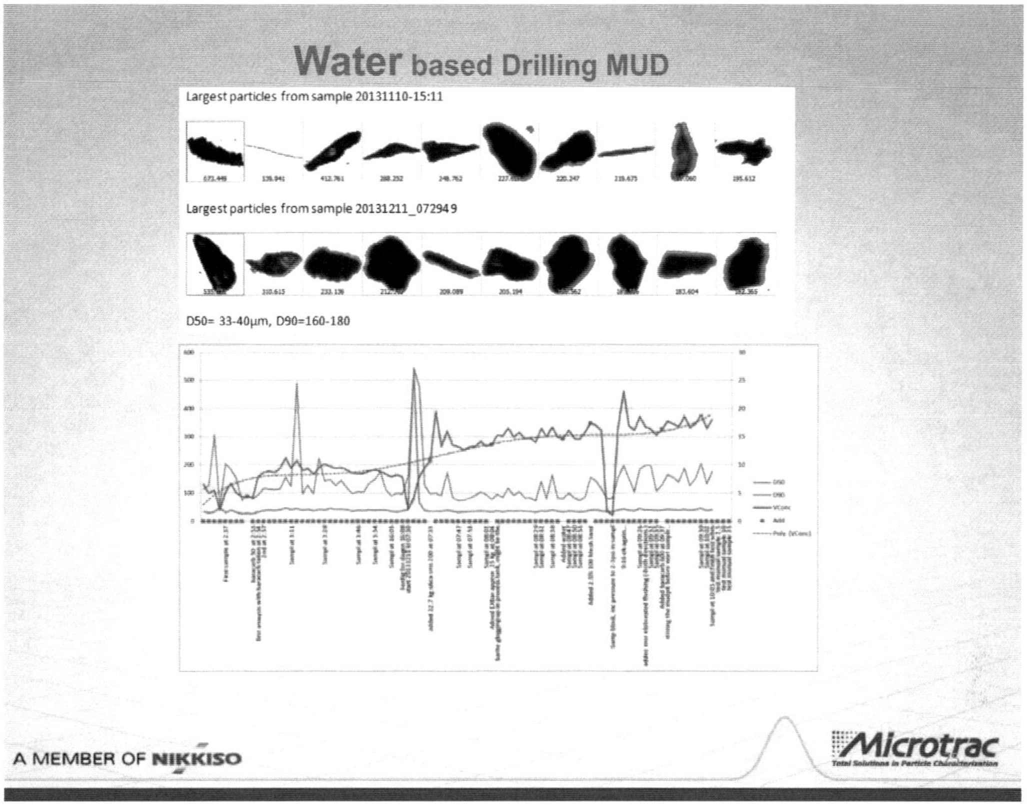

Water based Drilling MUD
Largest particles from sample 20131110-15:11
Largest particles from sample 20131211_072949
D50= 33-40μm, D90=160-180
A MEMBER OF NIKKISO
Microtrac
Total Solutions in Particle Characterization

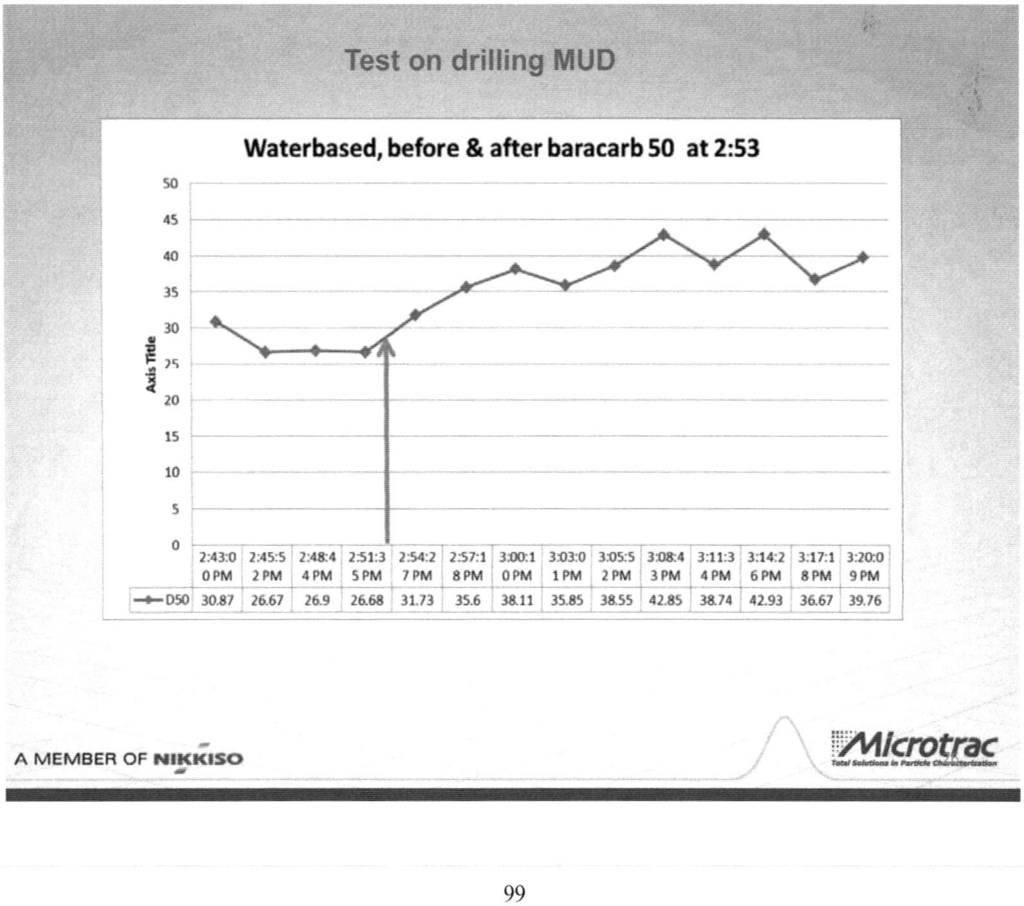

Test on drilling MUD
Waterbased, before & after baracarb 50 at 2:53
Axis Title
50
45
40
35
30
25
20
15
10
5
0
2:43:0 0 PM | 2:45:5 2 PM | 2:48:4 4 PM | 2:51:3 5 PM | 2:54:2 7 PM | 2:57:1 8 PM | 3:00:1 0 PM | 3:03:0 1 PM | 3:05:5 2 PM | 3:08:4 3 PM | 3:11:3 4 PM | 3:14:2 6 PM | 3:17:1 8 PM | 3:20:0 9 PM
D50 | 30.87 | 26.67 | 26.9 | 26.68 | 31.73 | 35.6 | 38.11 | 35.85 | 38.55 | 42.85 | 38.74 | 42.93 | 36.67 | 39.76
A MEMBER OF NIKKISO
Microtrac
Total Solutions in Particle Characterization

OIL based Drilling MUD

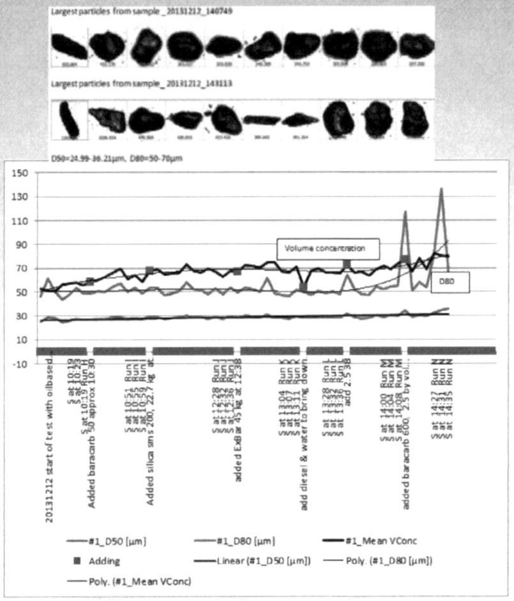

A MEMBER OF NIKKISO

Microtrac
Total Solutions in Particle Characterization

- Image analysis provides further characterization of particle systems.
- Full characterization of shape and form.
- 3-Dimensional imaging
- Fast, easy to use with complete software Windows interface.
- Dry powder and suspension measurements for LAB.
- Proven on-line dry and suspension measurements.

A MEMBER OF NIKKISO

Microtrac
Total Solutions in Particle Characterization

Thank You!
Questions ?
A MEMBER OF NIKKISO
Microtrac

Silicon Nanocrystals for Biological Imaging

Korgel, B. A., The University of Texas at Austin

Originally presented on: 4/30/2014 16:30:00 - 16:55:00

Silicon nanocrystals fluoresce with size-tunable color in the red-to-near infrared spectral range and are biocompatible, making them well-suited for in vivo biological imaging. There are however challenges facing their use in biological applications. Their band-edge absorption is relatively weak, leading to a large apparent Stokes shift between optical excitation and emission. To induce strong red light absorption while retaining the fluorescence properties of Si nanocrystals (i.e., red to NIR emission with long microsecond lifetimes), Si nanocrystals can be modified with molecular antenna with strong red absorption and charge transfer to the emitting nanocrystals. Si nanocrystals are also susceptible to oxidation and photoluminescence quenching in water, which has required ligand passivation chemistry utilizing bifunctional alkenes with ester or carboxylate terminal functional groups. Hydrophobic Si nanocrystals can also be dispersed in aqueous media using lipids and other surfactants. For specific targeting of molecular recognition groups on cells, the nanocrystals can be modified with either nucleic acids or peptides.

Capillary Torque on a Rolling Particle in the Presence of a Thin Liquid Film

Jeffrey S. Marshall
The University of Vermont

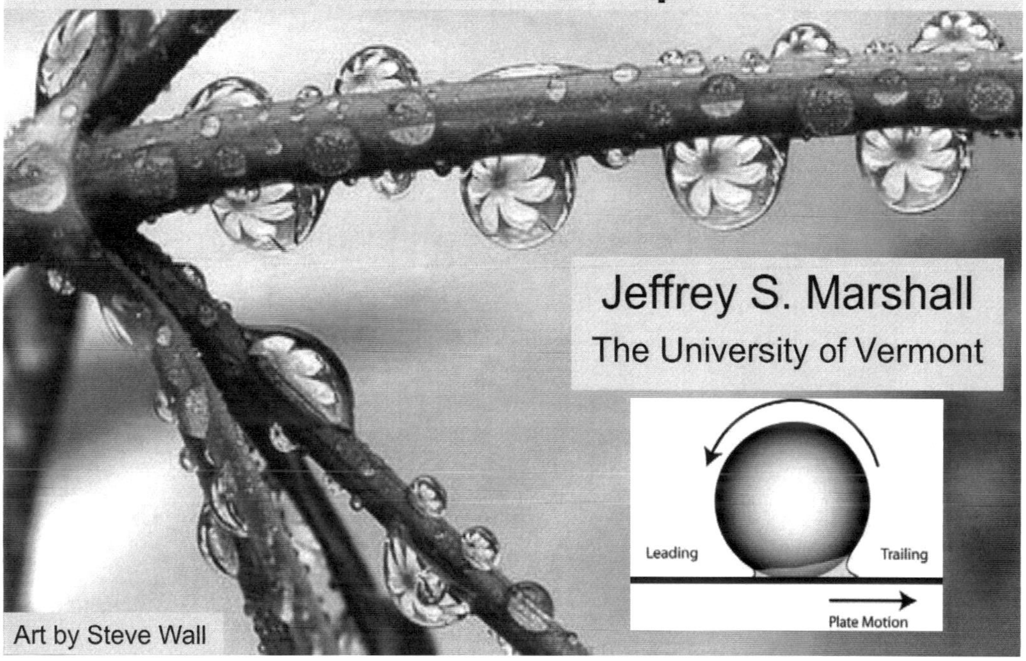

Liquid bridge force on particles

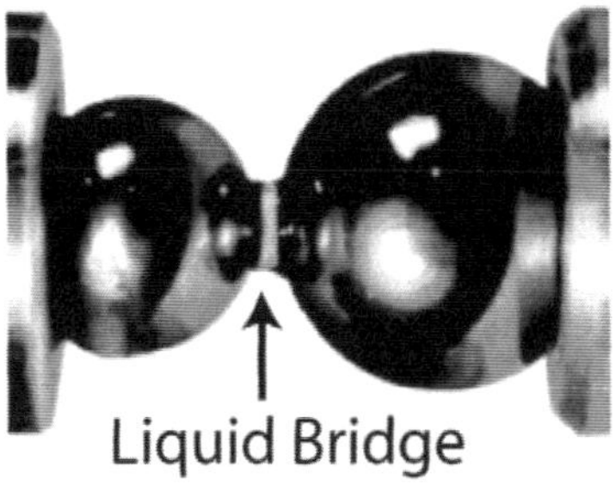

The presence of moisture gives rise to a capillary force that pulls particles toward each other. This can dramatically change the dynamics of granular flows, aerosols, etc.

Examples of wet granular flow

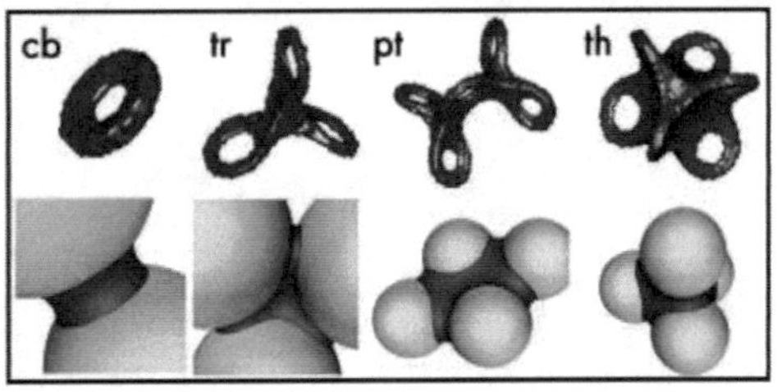

Mud slide

Normal capillary force

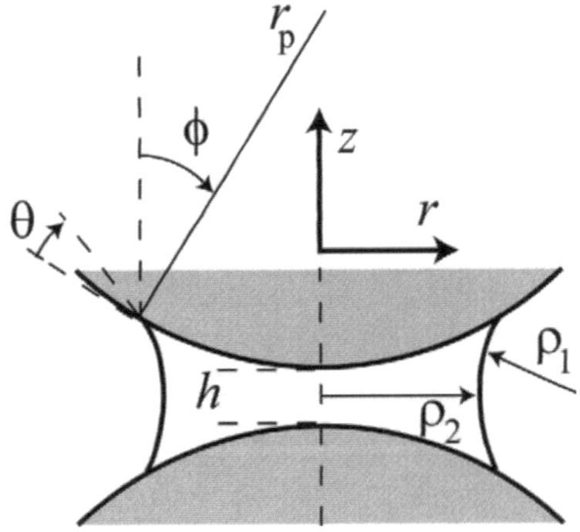

Normal Force on particles is caused both by:

- surface tension force at contact point
- pressure reduction within liquid bridge due to interface curvature PRIMARY CAUSE

$$F_{cap} = 2\pi\sigma\rho_2 + \pi\rho_2^2\Delta p_I = \pi\sigma\rho_2\left[1+\frac{\rho_2}{\rho_1}\right] \cong \pi\sigma\rho_2^2/\rho_1 \quad \text{for} \quad \rho_1/\rho_2 = O(\phi) << 1$$

Physical effects of particle rolling

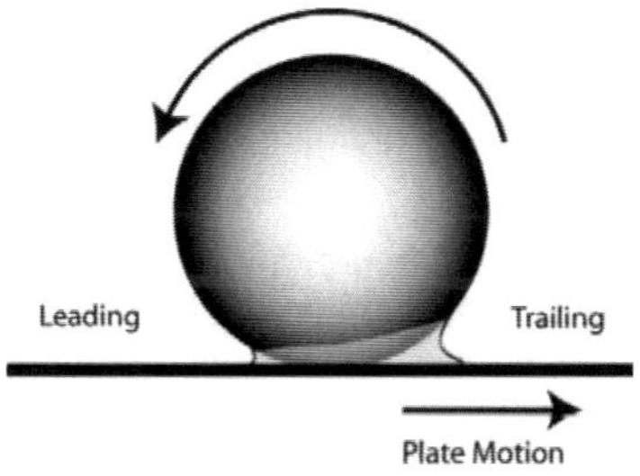

- liquid meniscus shifts toward trailing side of particle
- contact angles on two sides of particle become asymmetric

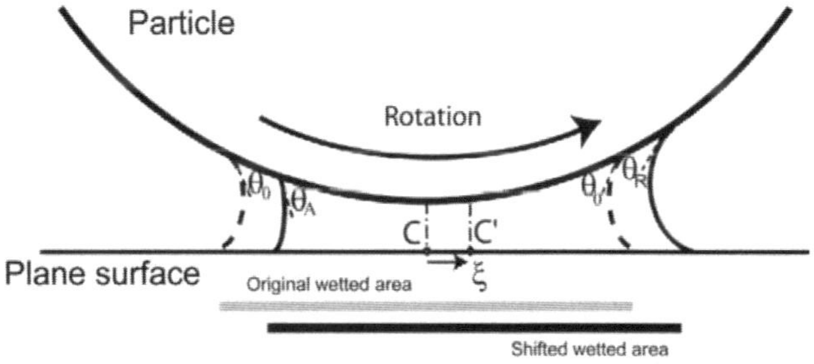

Two types of capillary torque

$$M_{pr} = \xi F_{pr}$$

$$M_{st} = 2\pi\sigma b r_p (\cos\theta_R - \cos\theta_A)$$

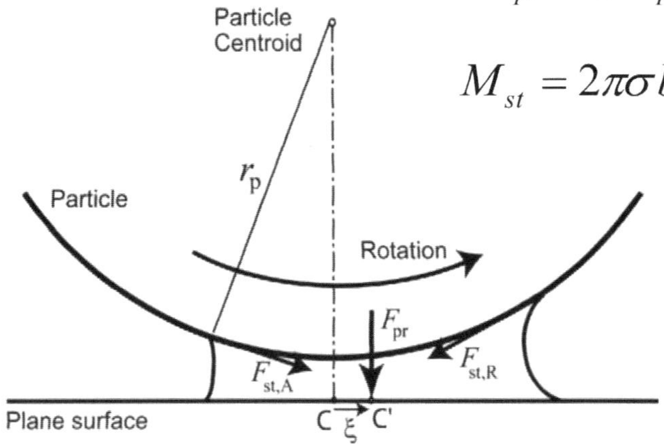

How to determine b, ξ, θ_A, and θ_R ?

Experiment – Bico et al (PF 2009)

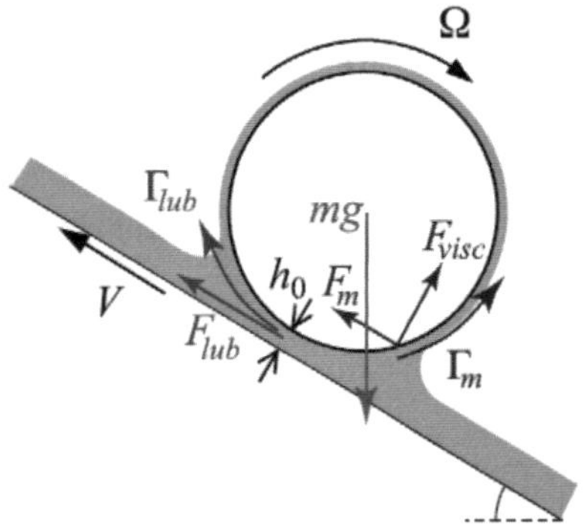

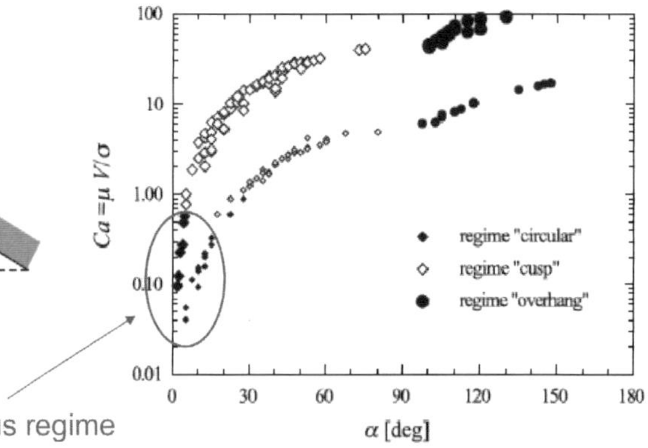

Experiment – Donahue et al (JFM 2012)

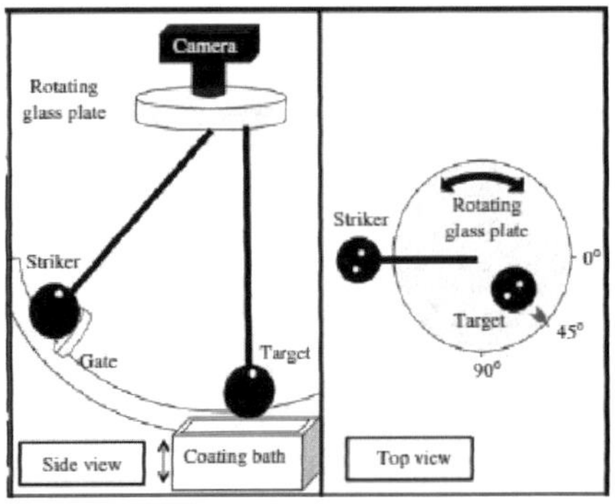

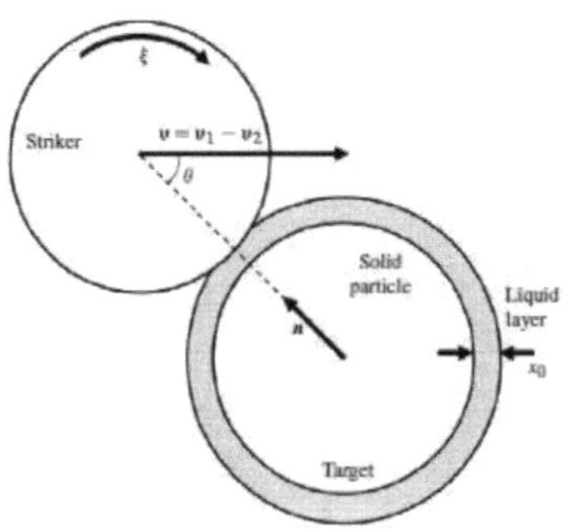

Measured restitution coefficient and receding contact angle for collision between two particles.

Experiment – Shade & Marshall (EF 2011)

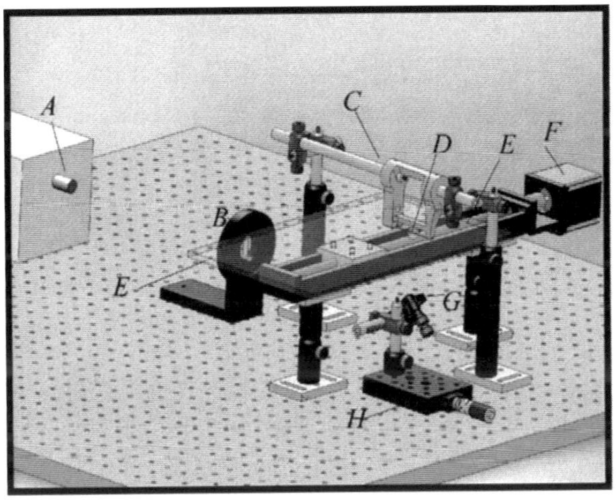

Experiment to measure contact angles and contact point shift for rolling particle with fixed centroid on translated plate.

Imaging

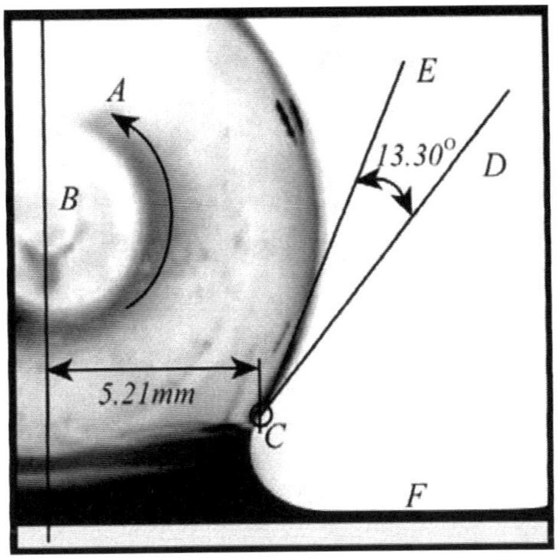

Contact angles and contact point locations measured using LIF.

Flow field within meniscus measured using PIV in vertical and horizontal planes.

Flow field at front in vertical plane

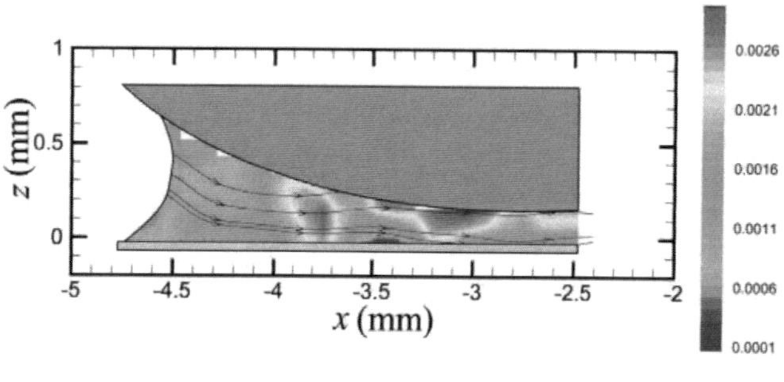

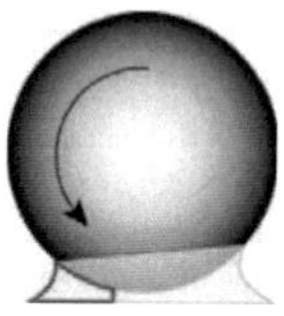

Flow field in rear in vertical plane

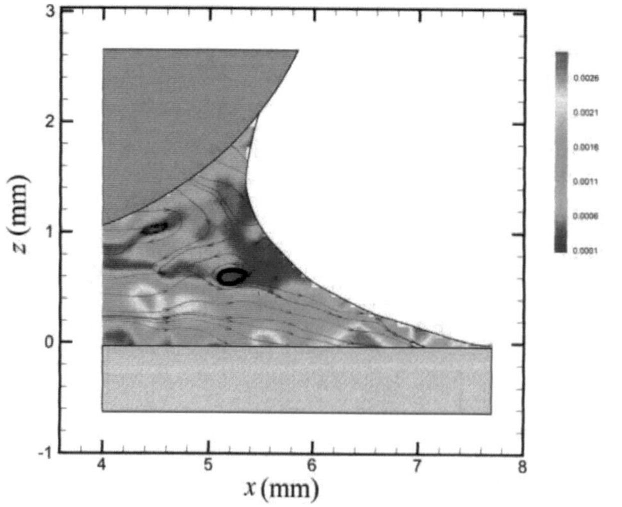

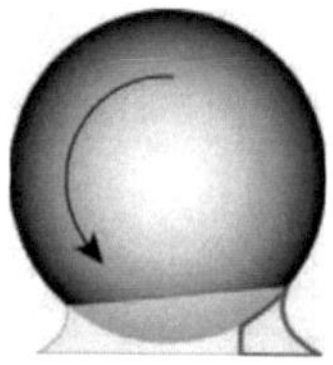

Flow in horizontal plane

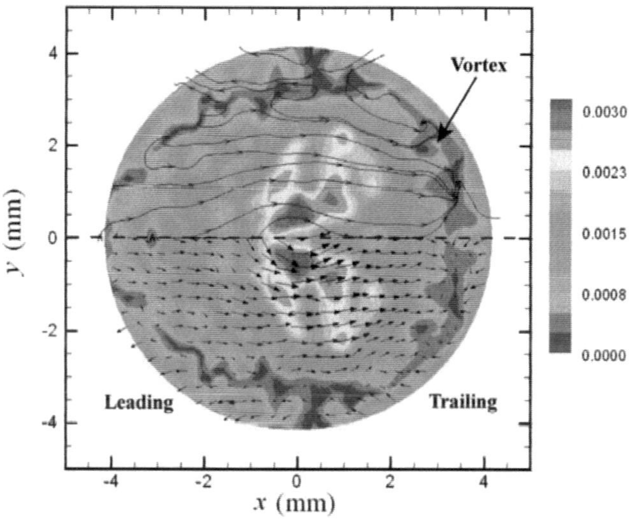

- Laser sheet is 1 mm above the glass surface.
- Colored region is liquid meniscus.

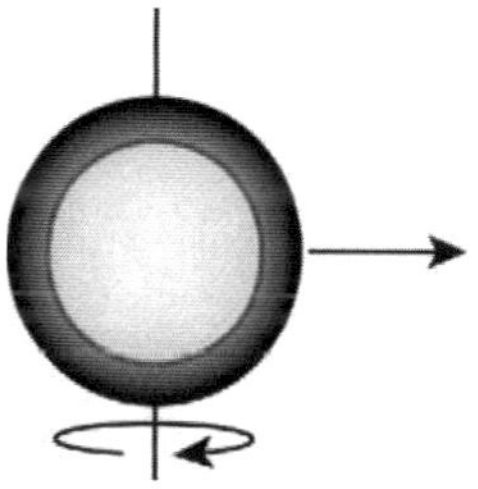

Data for rearward shift of liquid meniscus

Capillary number $Ca = \mu|V|/\sigma$

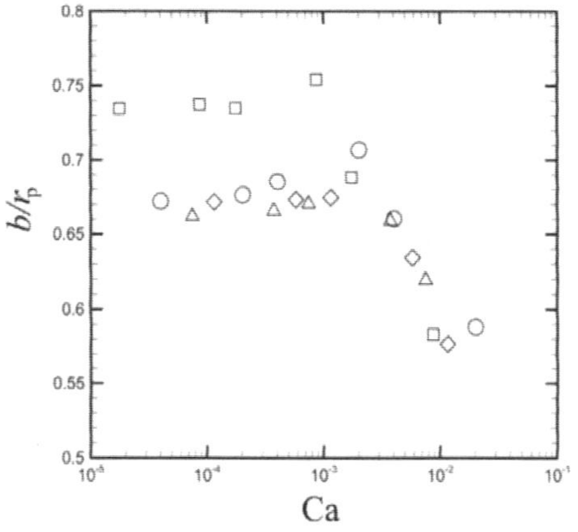

Average Contact Radius

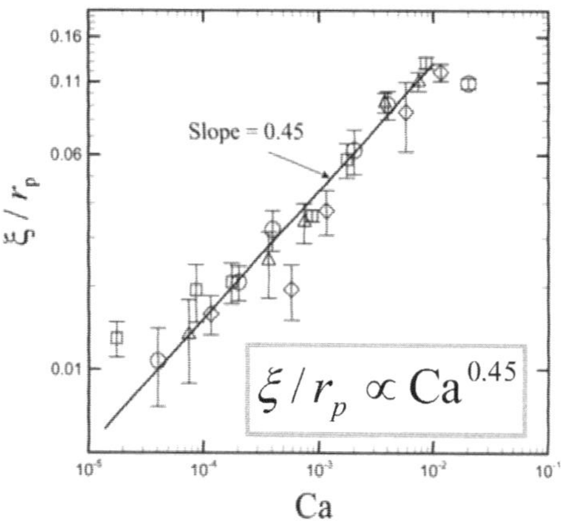

Rearward Shift of Liquid Meniscus

Data for receding and advancing contact angles

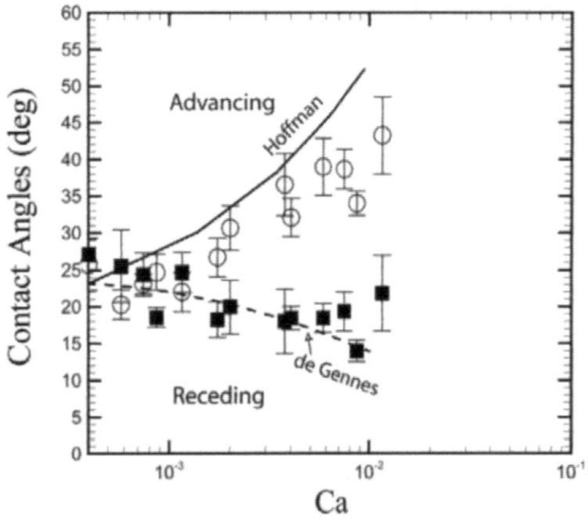

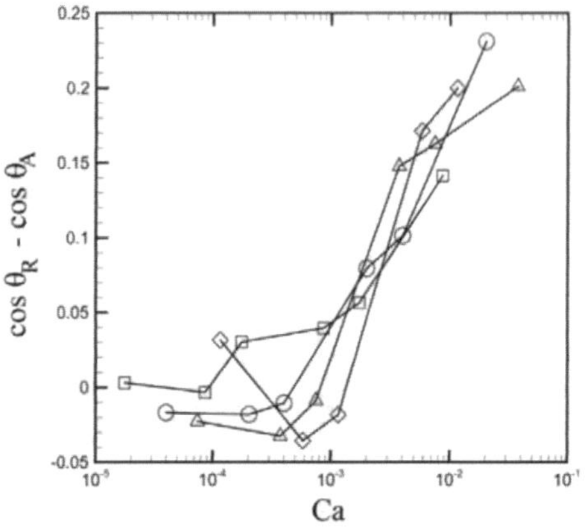

Dimensionless capillary torques

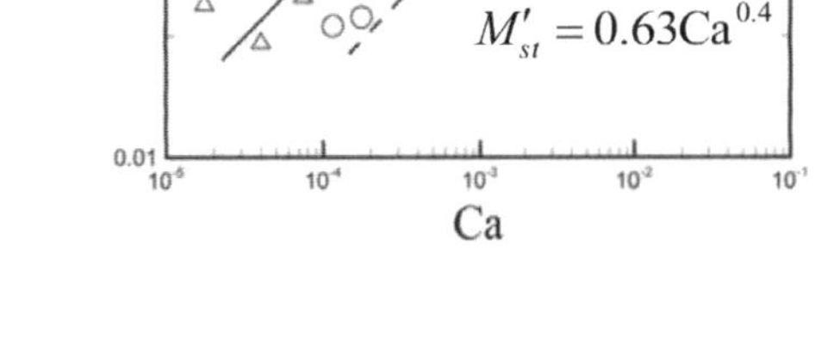

Torque due to shifted low pressure region:

$$M'_{pr} \equiv \frac{M_{pr}}{2\pi\sigma r_p^2} = (\xi / r_p)(\cos\theta_A + \cos\theta_R)$$

Torque due to asymmetric surface tension force:

$$M'_{st} \equiv \frac{M_{st}}{2\pi\sigma r_p^2} = (b / r_p)(\cos\theta_R - \cos\theta_A)$$

Viscous torque

Viscous torque acting on a particle rolling a distance δ above a wall at low Reynolds number in a homogeneous liquid with viscosity μ is given by

$$M_V = -\frac{4\pi}{5}\mu r_p^2 V\,[-3\ln(\delta/r_p)+5.7]$$

$$M_V = -8\pi\mu\, r_p^2 VA$$

Set δ based on surface roughness.

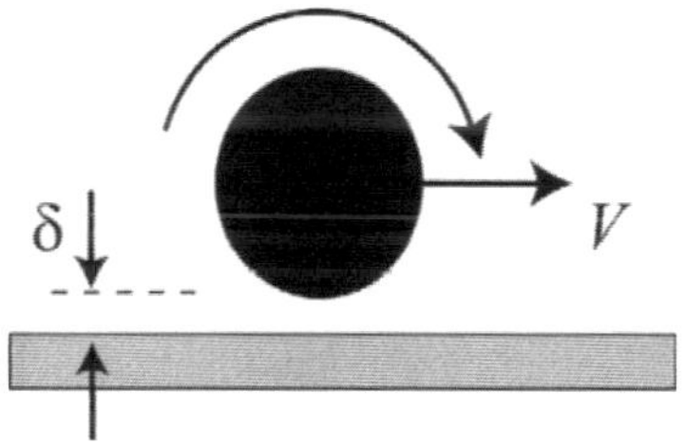

King and Leighton (1997)

Speed of particle rolling down an inclined plane with slope

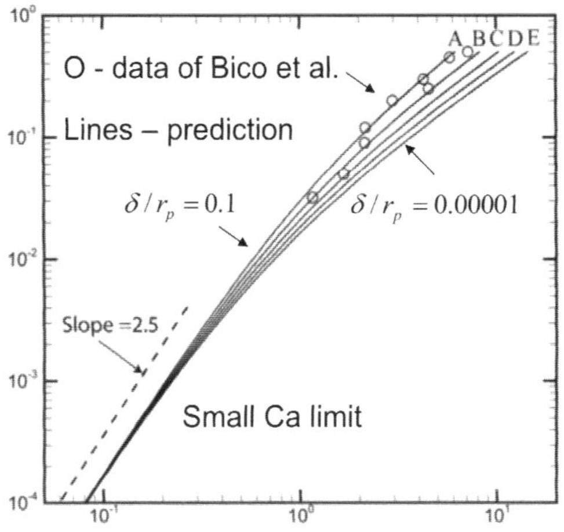

$$\mathrm{Bo} = \rho_p g r_p^2 \sin\alpha/\sigma$$

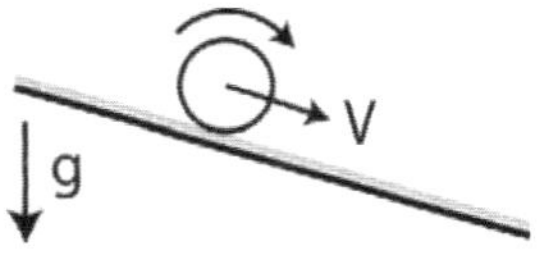

Equate torques due to gravity, viscous force and two capillary torques:

$$\frac{2}{3}\mathrm{Bo} = 4A\,\mathrm{Ca} + M'_{pr} + M'_{st}$$

Conclusions

- A particle rolling on an inclined plane in the presence of a liquid film experiences both a normal capillary force and a capillary torque.
- The capillary torque arises from both shifting of the low-pressure meniscus and asymmetry of the contact angles.
- We have developed a semi-empirical expression for capillary torque valid for low capillary numbers.
- The predictions agree closely with experimental data for a particle rolling down a wet slope.

References

Bico J, Ashmore-Chakrabarty J, McKinley GH, Stone HA. Rolling stones: The motion of a sphere down an inclined plane coated with a thin liquid film. *Physics of Fluids* **21**, 082103 (2009).

Donahue CM, Brewer WM, Davis RE, Hrenya CM. Agglomeration and de-agglomeration of rotating wet doublets. *Journal of Fluid Mechanics* **708**, 128-148 (2012).

King MR, Leighton DT. Measurement of the inertial lift on a moving sphere on contact with a plane wall in a shear flow. *Physics of Fluids* **9**(5), 1248-1255 (1997).

Marshall JS. Capillary torque on a rolling particle in the presence of a liquid film at small capillary numbers. *Chemical Engineering Science* **108**, 87-93 (2014).

Schade P, Marshall JS. Capillary effects on a particle rolling on a plane surface in the presence of a thin liquid film. *Experiments in Fluids* **51**(6), 1645-1655 (2011).

A little advertisement …

from Cambridge University Press

Modeling and Validation of the Cohesion Reduction of Fine Powders via Surface Modification

Dave, R. N., New Jersey Institute of Technology

Originally presented on: 4/30/2014 17:30:00 - 17:55:00

Flow and handling of fine powders is a topic of great interest to industry. High cohesion relative to their weight of fine powders leads to problems such as, agglomeration, poor flowability, electrostatic charging and low bulk density. In this talk, powders of interest to pharmaceutical industry are considered where poor flow problems lead to downstream problems such as poor content uniformity and marginal improvement in dissolution rate of fine API powders. To encounter these problems, nano-particle dry coating is shown to be a more rigorous approach. Surface modified pharmaceutical powders are produced using several devices, both based on batch and scalable continuous operation. Results illustrate the improvement in flow, fluidization, dispersion, lack of agglomeration, bulk density, and electrostatic tendency, all leading to potentially significant cost benefits in industrial processing, handling, storage and transportation. In order to understand the cause of cohesion reduction, contact modeling based on our multi-asperity advanced models where the influence of material properties, surface area coverage and spatial and size distribution of guest particles is presented. Models are qualitatively validated by bulk and particle scale experimental results, including detailed surface energy measurements that explain cohesion of pharmaceutical powders based on two major tuning parameters, surface roughness and surface energy. Issues of linking particle to bulk scale are discussed along with selected results from Discrete Element Modeling of cohesive powders.

Engineered Particle Systems for Inhaled Drug Delivery

Maynor, B., Liquidia Technologies

Originally presented on: 4/30/2014 18:00:00 - 18:25:00

Particle properties such as size, shape, and density play a critical role in determining the performance of inhaled medicines. Here, we will highlight progress in the field of "engineered particles" for pulmonary delivery and highlight the use of a particle molding technique called PRINT® (Particle Replication in Non-wetting Templates) (Liquidia Technologies, Inc). PRINT molding produces precisely controlled particles on the micron and nanometer scale for drug delivery. Unlike other particle generation technologies, the PRINT molding technique simultaneously enables precise control over particle size, shape, and chemical composition. Examples from the application of the PRINT technology and other "engineered particle" generation approaches towards advancing the field of pulmonary medicine will be described here, including improvement of particle dispersion, dissolution enhancement of poorly soluble compounds, and control of cellular interactions.

Electrostatic Charging and Entrainment Behaviors of Binary and Quaternary Particulate Systems in Fluidized Beds

Zhou, X. H., Columbia University

Originally presented on: 4/30/2014 18:30:00 - 18:35:00

A number of commercial fluidization processes involve binary or quaternary particulate systems with different particle sizes and densities. During the operation of multiphase systems such as fluidized beds, electrostatic charges are generated primarily via triboelectric or frictional charging due to the dielectric nature of the materials. The accumulation of electrostatic charge within the system can impact the fluidization behavior and in some cases can be operationally hazardous. In this study, the electrostatic charge generation and accumulation are investigated for binary and quaternary particulate systems using a faraday cup system and an on-line electrostatic probe system. Specifically, the effect of addition of two different fine iron ores (i.e., hematite and magnetite) in fluidized beds is studied in terms of particle-particle interactions and entrainment rates. The behaviors of different particulate systems are found to highly depend on the chemical and physical properties of particles such as size, density, hydrophobicity, surface roughness and even magnetism. The results suggest that the enhanced electrostatic forces between fine and coarse particles due to significant electrostatic charging phenomenon during fluidized bed operation can retain the fines to some extent.

Vibrations of Jammed Disk Packings with Hertzian Interactions

Schreck, C., Lawrence Berkeley National Laboratory

Originally presented on: 4/30/2014 18:48:00 - 18:53:00

Granular media has complex vibrational properties. Contact breaking and Hertzian interactions between grains can both give rise to nonlinear vibrational response of static granular packings. Through molecular dynamics simulations in model granular packings that interact via Hertzian spring potentials, we show that while Hertzian interactions lead to weak nonlinearities in the vibrational behavior (e.g. the generation of harmonics of the eigenfrequencies of the dynamical matrix), the vibrational response of static packings with Hertzian contact interactions is dominated by contact breaking as found for systems with repulsive linear spring interactions.

Computed Tomography Study of Internal Structure of 3D Granular Packing based on Shape

Murphy, K., University of Chicago
Jaeger, H., University of Chicago
Originally presented on: 4/30/2014 18:54:00 - 18:59:00

One of the aims of granular physics is to understand how local interactions between macro-particles can give rise to large scale properties of the aggregate. We investigate such local interactions when external mechanical stresses are applied. X-ray computed tomography is used to monitor and track local (re-)configurations inside three-dimensional aggregates of 3D-printed particles of various shapes. By comparing aggregate reconstructions at various strain values, from no applied stress to just after the yield point where the granular packing transitions into the plastic regime, we examine the dependence on particle shape of the local particle reconfigurations that are responsible for yielding or occur as precursors to failure.

Numerical Simulations of Fluidized Suspensions of Cohesive Particles with Size Distributions

Ozel, A., Department of Chemical and Bological Engineering, Princeton University
Gu, Y., Rose-Hulman Institute of Technology
Radl, S., Graz University of Technology
Sundaresan, S., Princeton University
Originally presented on: 4/30/2014 19:00:00 - 19:05:00

Euler-Lagrange simulations of fluidized beds, where the locally-averaged equations of motion for the fluid phase is solved in an Eulerian framework and the particles are tracked in a Lagrangian fashion, have found widespread use in the study of fluidization [1-4]. This approach is commonly referred to as CFD (computational fluid dynamics)-DEM (discrete element method) when all the particles are followed and as CFD-DPM (discrete parcel method) or MP-PIC (multiphase particle-in-cell) when only a small number of representative particles (a.k.a. "parcels") are simulated [3,4]. Many research groups are currently investigating the effects of cohesion and size distribution on fluidization characteristics via CFD-DEM simulations [for example, see ref. 1, 2].

We have performed CFD-DEM simulations of gas-fluidization of Geldart Group A particles in small periodic domains in order to probe the fluidization characteristics of gas-fluidized suspensions at volume fractions typically observed in circulating fluidized beds and turbulent fluidized beds. In this poster, we will present results related to the following three questions: (a) How does the inclusion of cohesion alter the fluidization of monodisperse particles? (b) How does the addition of fines affect fluidization when cohesion is not considered? (c) What is the combined effect of fines and cohesion?We also ask how one should systematically coarse-grain the results of such detailed CFD-DEM simulations to obtain effective constitutive models (for example, for the fluid-particle interaction force) that are needed for affordable simulations of industrial scale beds, which entails coarse fluid grids as well simulation of only a small number of representative particles. As shown previously [5,6], when such coarse-graining is done, the drag coefficients must be corrected. We will demonstrate in this poster that both coarsening the fluid grid and switching from DEM to DPM lead to lower effective drag coefficients.

Shear Thickening of Chemical Mechanical Polishing Slurries

Krasovsky, A., Department of Chemical and Biological Engineering, Colorado School of Mines

Originally presented on: 4/30/2014 19:06:00 - 19:11:00

Numerous YouTube videos show people walking on "water". This "water" is a mixture of cornstarch and water that exhibits a phenomenon known as shear thickening. The viscosity of shear thickening fluids dramatically increases upon reaching a critical stress (e.g., when a human foot contacts the fluid), but behaves like a liquid below the thickening threshold. Chemical mechanical polishing (CMP) slurries also exhibit shear thickening. CMP is a fundamental technology used in the semiconductor manufacturing industry to polish and planarize a wide range of materials for the fabrication of microelectronic devices (i.e., computer chips). During the high shear polishing process, it is hypothesized that individual slurry particles (~100 nm) collide with one another to form large agglomerates (>500 nm) that cause the slurry to shear thicken. These agglomerates tend to dig into the material surface triggering defects such as scratches or gouges during polishing (costing the semiconductor industry billions of dollars annually). The project's goal is to understand the high shear rheological behavior of CMP slurries and to link changes in particle structure, both temporary and permanent, to the observed shear thickening response through the use of high shear (>10,000 s-1) rheology. Small-angle light scattering was employed in situ with rheological characterization (rheo-SALS) to identify the formation, shape, and size of agglomerates generated under shear. Current work is transitioning from studying fractal shaped fumed silica (currently used by industry) to spherical colloidal silica nanoparticles.

Keplerian Orbits in Freely Falling Charged Grains

Lee, V., University of Chicago

Originally presented on: 4/30/2014 19:12:00 - 19:17:00

We report high-speed tracking of freely falling charged grains in the co-moving frame. Elliptical and hyperbolic Keplerian orbits due to electrostatic interaction between oppositely charged grains are observed directly. By using Nelder-Mead optimization method, we are able to reconstruct the Keplerian trajectories in three-dimensional space. We also find that electrostatic force can lead to aggregation in a granular stream.

A Particle Segregation Study Using GPU-Accelerated DEM Simulations with Elastic Tangential Friction

Isner, A., University of Kentucky
Umbanhowar, P. B., Northwestern University
Ottino, J. M., Northwestern University
Lueptow, R. M., Northwestern University
Originally presented on: 4/30/2014 19:18:00 - 19:23:00

The discrete element method (DEM) is a popular simulation technique that has proven useful in the study of granular flows, particularly their assembly microstructure, interparticle force networks, stress distribution, and flow kinematics. Of practical concern is the simulation of particle systems at industrially relevant length and time scales, which may require simulating tens to hundreds of millions of particles using a variety of flow geometries and contact force models. Although progress has allowed these computationally expensive simulations to run on highly scalable multi-processor systems, this approach may require significant capital investment or access to high performance computing resources. However, general purpose graphics processing unit (GPU) computing approaches have recently become a viable alternative that leverages the built-in parallelism of discrete GPU hardware, which can contain many thousands of parallel processing cores on a single device, as opposed to the four or six cores found in many modern CPU's. Here, we present a proof-of-concept implementation of DEM on a consumer-grade GPU based on Nvidia's CUDA Particles SDK, which incorporates a linear spring dashpot normal force model and a sliding and elastic tangential frictional force model. The DEM simulations are validated by comparing particle streamwise concentration profiles in quasi-2D bounded heap flows of bidisperse mixtures to those of experiments. We investigated the dependence of several kinematic parameters including flowing layer depth, diffusivity, and percolation length on various control parameters such as particle size ratio and two-dimensional feed flow rate. In particular, we find that the interparticle percolation length scale normalized by the small particle diameter is approximately proportional to the logarithm of particle size ratio, R, (R = dl / ds, where dl and ds are the diameters of the large and small particle, respectively) for size ratios in the range R = 1.1 to R = 3. An approach to generalize our understanding of shear-driven segregation to polydisperse systems is also tested using GPU-accelerated DEM simulations of polydisperse mixtures. As well as achieving real-time visualization of simulation results, we achieve a performance improvement that is on average one order-of-magnitude greater than an analogous DEM algorithm run on a traditional multi-core processor. The cost-savings (on a TeraFLOPS/$ basis) of GPU-accelerated DEM makes GPU-based DEM a promising platform for sustained scalability in simulations of certain granular systems.

Subsurface X-ray Visualization of Three-Dimensional Granular Tumbler Flow and Mixing

Zaman, Z., Northwestern University
Umbanhowar, P. B., Northwestern University
Ottino, J. M., Northwestern University
Lueptow, R. M., Northwestern University
Originally presented on: 4/30/2014 19:24:00 - 19:29:00

Granular flow and mixing experiments in rotating partially filled tumblers are often limited to visualization along walls or the free surface. To probe the dynamics deep inside the bed, a common approach is to experimentally utilize quasi-two-dimensional (quasi-2D) geometries and infer the three dimensional (3D) behavior by assuming that the flow is primarily 2D with diffusive spanwise particle motion. However, this model can miss subtle effects seen solely in 3D geometries. For example, we demonstrate that an axial drift of monodisperse particles exists in partially filled spherical and double cone tumblers using experiments and discrete element method (DEM) simulations. The drift is small (1-2 particle diameters or 1-3% of the tumbler diameter per flowing layer pass) relative to streamwise displacements, whose speed depends on the flowing layer length.

To further our understanding of 3D flow and mixing in tumblers, we designed and built an X-ray imaging system that provides subsurface imaging of the tumbler during flow. Tracer (X-ray opaque) particles in a bed of X-ray translucent particles are tracked to construct trajectories through the particle bed. Stroboscopic imaging of the tumbler can provide Poincaré sections to determine regions of good mixing and regions of segregation. Current work involves providing experimental support for the existence of persistent segregated regions for a two axis tumbling protocol, as predicted by a dynamical systems theoretical model. Ultimately, the verification of the model will allow for informed design of tumbling protocols. Funded by NSF Grant CMMI-1000469.

An Industrial Perspective on Problems facing the Particle Technology Community

Karl Jacob
The Dow Chemical Company
Engineering & Process Sciences
Solids Processing Lab
Midland, Michigan
April 30, 2014

Outline for today

- Opening remarks
- Significant improvements in particle technology/solids processing over the last 25 years
- Challenges we face as a scientific community going forward
- Some final thoughts...

Do bulk solids cause problems?

- Wrong particle size/shape/characteristics
- Equipment plugged/jammed
- Rate through plant is less than designed
- Chemical degradation of product

Just a FEW of the issues that are observed when plants handle solids

Intent/Disclaimer

- Attempt to highlight some of the industrial and scientific advances over the last quarter of century which I feel have greatly impacted the field
- Not meant to be a complete review of the literature over the last 25 years – that would be impossible even in two days!
- For any sins of omission that I commit over the next 30-40 minutes, I ask in advance for your pardon and forgiveness

Why 25 years?

- Dow Solids Processing Lab 25 years old on 1/20/14
- One senior consultant said, "the average life of an industrial solids lab is 10 years".
- The people and the problems are reasons to come to work each day!

Keep your day jobs folks!!!!

- As a community we continue to impact society in both large and small ways such that it is very important to continue to make particle technology more accepted in the engineering sciences
- Hunter
- AIChE North Central Student meeting – Michigan State University – April 12

Solids Processing Technologies

- ***Agglomeration***
- ***Storage/Powder Mechanics***
- Pneumatic Conveying
- Mechanical Conveying
- Fluidization
- Drying
- Particle Characterization
- Particle Formation
- Solid/Liquid Separation
- Solid/Gas Separation
- Classification/Screening
- ***Size Reduction***
- Feeding and Metering of Solids
- Dust Explosions/Process Safety
- Mixing and Blending
- Particle Engineering

Quantification of Segregation

- **Prior science**
  - **Broad heuristics**
  - **Knew mechanisms but difficult to quantify**
  - **Considerable work done by Shinohara at Hokkaido U.**
- **Program at Northwestern started in 2010 with Rich Lueptow and Julio Ottino**

Regime map for segregation

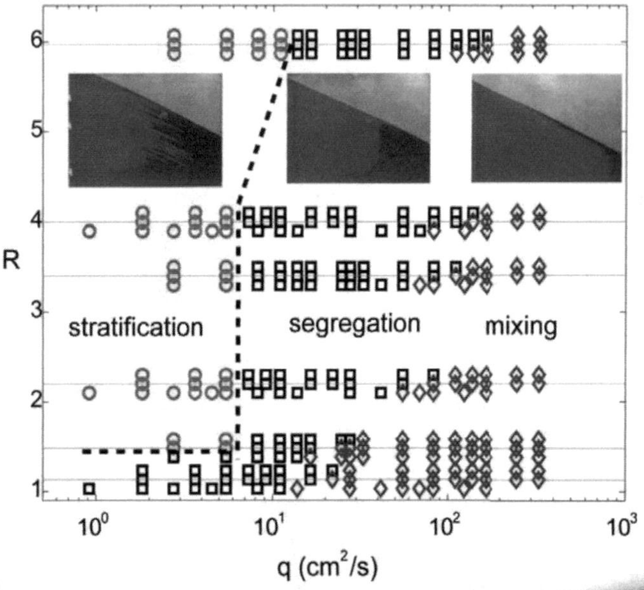

- T = 1.27 cm
- W = 46, 69, 91 cm

9

Three final particle configurations

Stratified | Segregated | Mixed

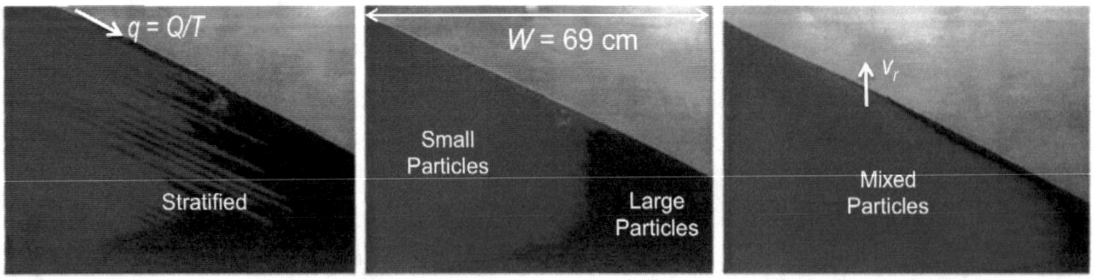

Q = 1 cm^3/s | Q = 24 cm^3/s | Q = 420 cm^3/s

- Size ratio $R = d_l / d_s = 3.4$
- Blue: d_l = 1.69 mm
- Red: d_s = 0.5 mm

Fan et al. (2012) Physical Review E

Governing equations

Transport equation for the volume concentration of each species c

$$\underbrace{\frac{\partial c}{\partial t}+u\frac{\partial c}{\partial x}+w\frac{\partial c}{\partial z}}_{\text{Advection}}+\underbrace{\frac{\partial}{\partial z}\left(c\boxed{v_p}\right)}_{\text{Segregation}}=\underbrace{\frac{\partial}{\partial z}\left(\boxed{D}\frac{\partial c}{\partial z}\right)}_{\text{Diffusion}}$$

What we need to know:

- Percolation velocity v_p
- Diffusion coefficient D

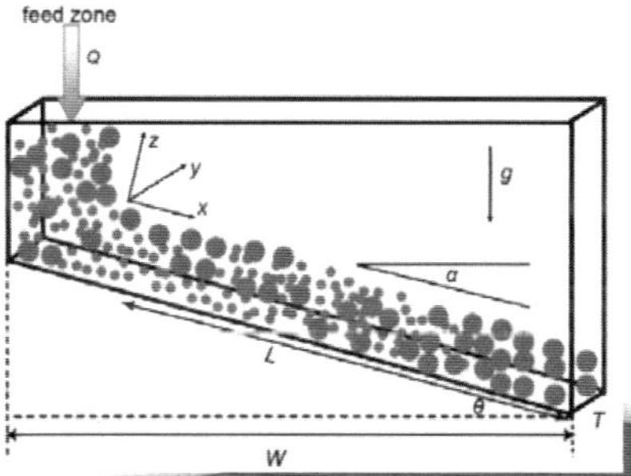

Percolation velocity and diffusion coefficient

$$\frac{\partial c}{\partial t}+u\frac{\partial c}{\partial x}+w\frac{\partial c}{\partial z}+\frac{\partial}{\partial z}\left(c\boxed{v_p}\right)=\frac{\partial}{\partial z}\left(\boxed{D}\frac{\partial c}{\partial z}\right)$$

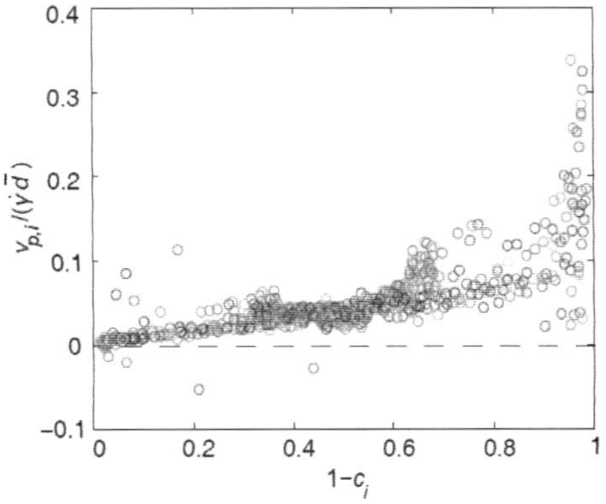

$v_{p,i} = w_i - w$

- **Percolation velocity:**

$$v_{p,i} = S(1-c_i)\dot{\gamma}$$

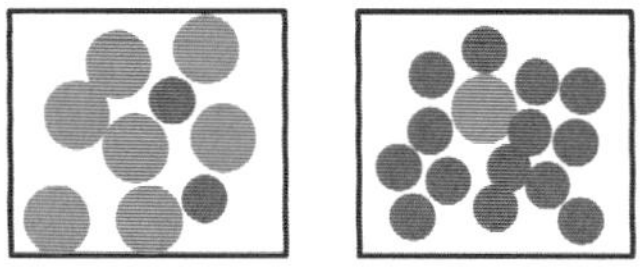

- **Diffusion coefficient:** Constant value measured from DEM simulations

Comparison with simulation and experiment

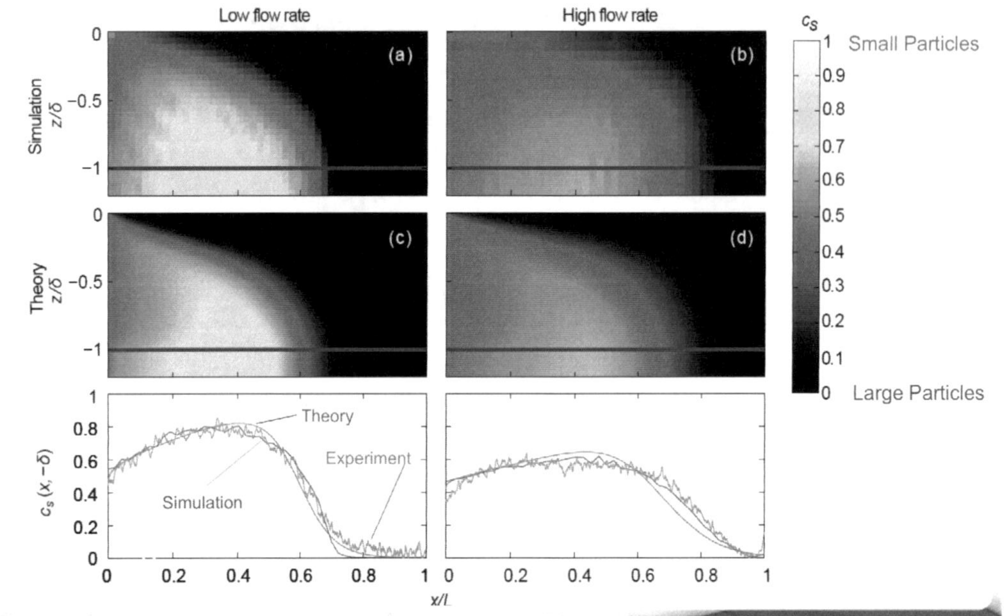

Fan et al., to be submitted to the Journal of Fluid Mechanics

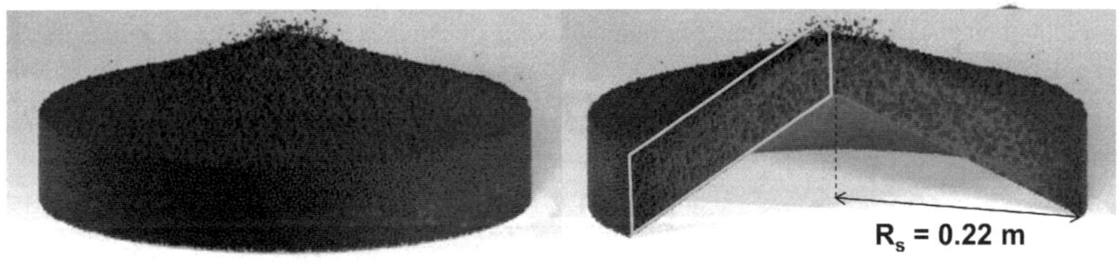

- ~120 g/s feed rate
- 2 & 4 mm particles
- 0.22 m silo radius
- Same force model parameters as used in quasi-2D heap simulations (ρ = 2.5 g/cm^3)
- Simulated time – 180 seconds
- Final particle count – 1.15 million
- Real time elapsed – 3 days

Rough comparison of theory to DEM simulation

Concentration at bottom of flowing layer

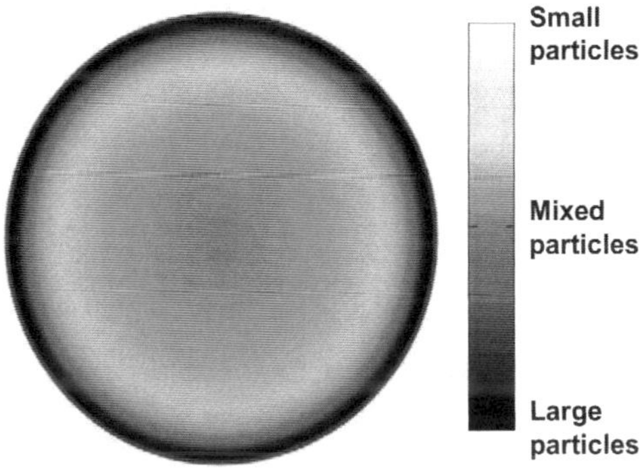

3D DEM simulations to calculate the flow kinematics

In both cases, particles remain mostly mixed upstream, and there is a band of just large particles at the outer edge of the silo

Jenike Tester

Development of the RST by Dietmar Schulze

Ring Shear Tester – Serial No. 1
Delivered to Dow Stade Solids
Processing Lab in early nineties

Advantages of RST

- Productivity – One could get twice the data in one quarter of the time for well-behaved materials
- Eliminated pre-consolidation step
- Infinite travel made testing of "difficult" solids considerably easier
- RST-XS made glove box testing of air or moisture sensitive solids easier

- Consequences of RST development
  - RST became more of a research tool
  - More likely to do shear testing
  - Students routinely use the RST as part of their theses work to quantify flow

Computer-based tools

- Tremendous tool development in the last 25 years
  - Solidsim process flowsheeting (now part of AspenTech) by Werther et al
  - Discrete element modeling
  - CPFD Barracuda®
  - Fluid/solid coupling in commercial CFD packages
  - Population balance modeling
- Enabled us to "see" inside our vessels, understand process dynamics, understand the implications of design choices, etc.
- Some cautions
  - Validation required
  - One size doesn't fit all
  - Not all problems require using these tools
  - Need to insure practitioners understand the particle technology not just the program

Submicron/Nanoparticle Fine Grinding

- **Single micron particle sizes produced by jet milling or an attritor mill was typically the lower limit for mechanical grinding**
- **Improvements in fine YTZ beads, bead separation technology and wear components in stirred media mills have enabled fine grinding down by mechanical means into the nanometer range**
- **Considerable body of work done by Schwedes, Kwade, Mende, et al. at TU Braunschweig**
- **Numerous applications of this technology**
  - **Energy storage devices**
  - **Pharmaceuticals**
  - **Paints/pigments**
  - **Many more to come...**

Typical Bead Mill

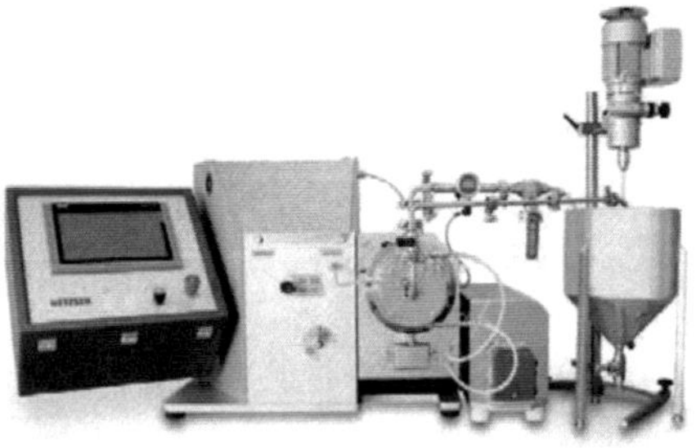

Netzsch LabStar®

Typical mill agitator/body arrangements

From Netzsch website

Typical particle size from fine media milling

Influence of the Grinding Media Size

NETZSCH

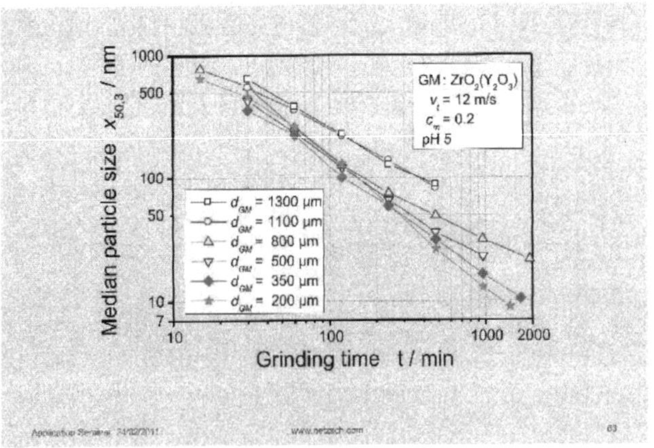

Fine Grinding Challenges

- Significant increase in surface area means particle interaction physical chemistry will be very important
- Understanding suspension stability is an absolute must for both aqueous and organic systems
- Understanding viscosity vs. size vs. stability key to successful milling

- To this point in time, everybody is content that we can just get to the nano range
- Classification of various fractions or removing the oversize or superfine fractions will be the key to successful application of these products in the marketplace

Just a few of the challenges that lay ahead of us

- Powder flow
- Dense phase conveying
- Fundamental particle/particle and particle surface interactions
- Education

Powder Flow Challenges

- Accurate rathole prediction
- Transmission of waves in bulk solids
- Sticky/elastic solids
- Caking/Blocking/time consolidation

For more detail, see "Opportunities and Challenges in Bulk Solids Handling" by John W. Carson of Jenike & Johanson

Dense phase conveying

- Some very nice progress here over the last 25 years, especially recent work by Kalman and Klinzing
- Standard paradigm is still to test at small full scale and use scaling heuristics to scale to full scale design
- Both Waeschle Coperion and Zeppelin have done considerable work in both understanding and designing large scale systems (6 inches or greater pipe diameter) to resist dynamic forces associated with directional changes of conveying slugs

Funnel Flow

- Approaches thus far
  - Jenike ff=1.7
  - Johanson, JR, effective head approach
  - Arnold and Roberts, average the above
  - Schwedes

- So what do we do?
  - experienced practitioners know what bound to choose
  - consider the economic consequence of a 6 ft outlet size

Very simple question: how big should the hole be to prevent the formation of a stable rathole?

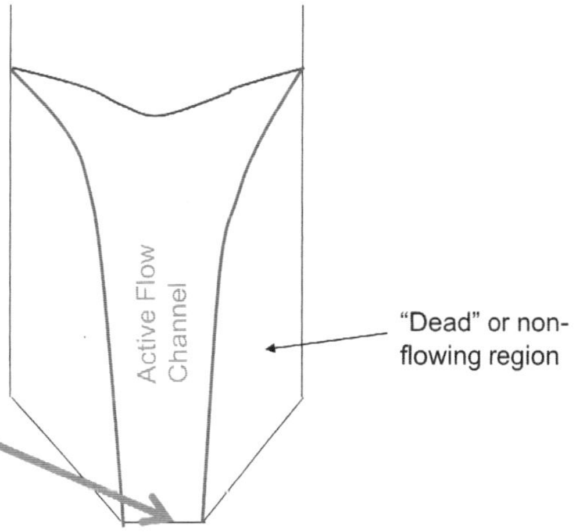

Understanding Pressure Waves/Vibration through Bulk Solids

- Air Cannons and vibrators used routine to "promote" flow
- All design paradigms are empirical
- Need to understand application of these devices vs. the condition of the material in the bin
- VERY little work done in the area

Fundamental Particle/Particle & Particle Surface Interactions

- Electrostatics – while considerable progress has been made by Dan Lacks of Case Western Reserve U., it remains a difficult area experimentally and vexing production problem
- Friction – still don't understand it – extraordinarily difficult to make a priori predictions about friction between surfaces and bulk solids – mill sludge
- Stick slip phenomena

Corn Starch – Stick/Slip

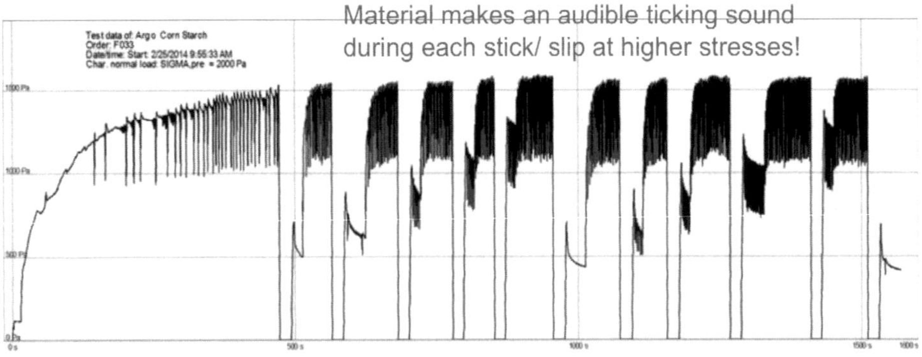

What is it about the particle size, PSD, processing history that results in this stick slip behavior?
What does the frequency of the stick/slip tell us about the particle/particle interactions?
Could this be predicted a priori?

Time Consolidation/Caking

- Very difficult, but very ubiquitous problem
- Depends greatly on mechanism
- Think about the following
  - Inorganic halides and salts, Epsom salts at Walmart
  - Blocking/massing of polymers
  - Salicylic acid
- Need to think along the lines of Kerry's approach yesterday

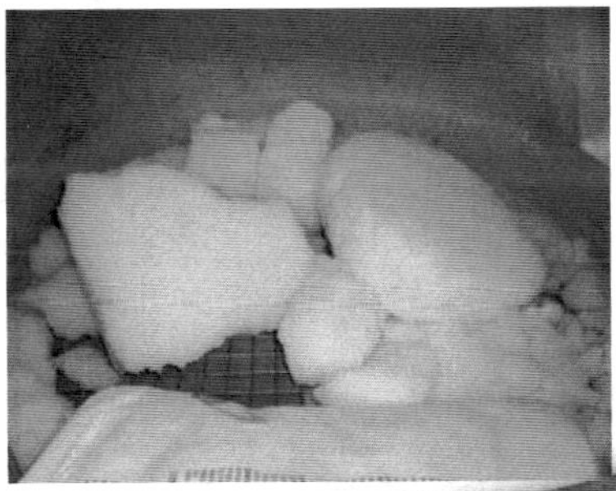

Education is key!

- **Still only a few programs have a technical engineering elective in solids processing/particle technology**
- **New program at Delaware spearheaded by Norm Wagner/Jim Micheals**
- **Programs at Akron, Purdue, NJIT, Rutgers, CCNY, Texas A&M, IIT, ...**
- **Progress with webinars, short courses, postgraduate training**
- **Plenary session at AIChE in San Francisco highlights the gap between industry and academia – see editorial by C. Moscone in April CEP**

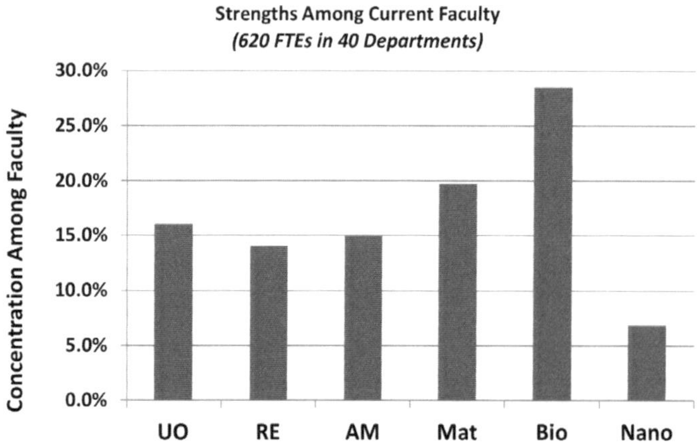

	FTE Concentration	Relative Strength
Bio	28%	5
Mat	20%	4
Particles!! → UO	16%	3
AM	15%	2
RE	14%	2
Nano	7%	1

33

Evolutionary Trend ?

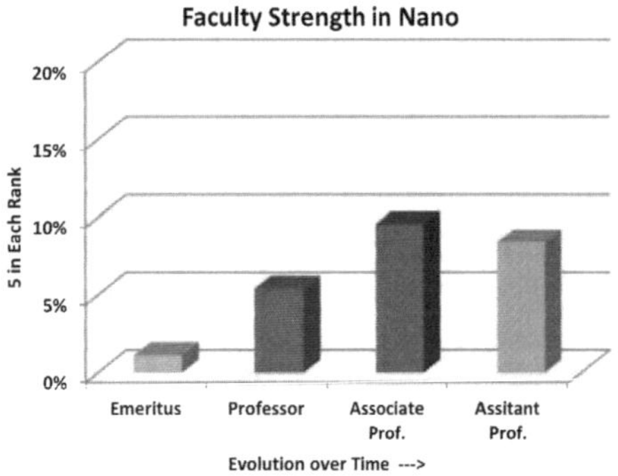

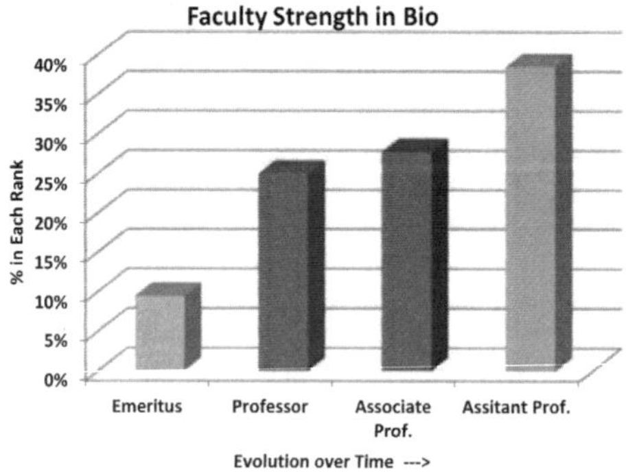

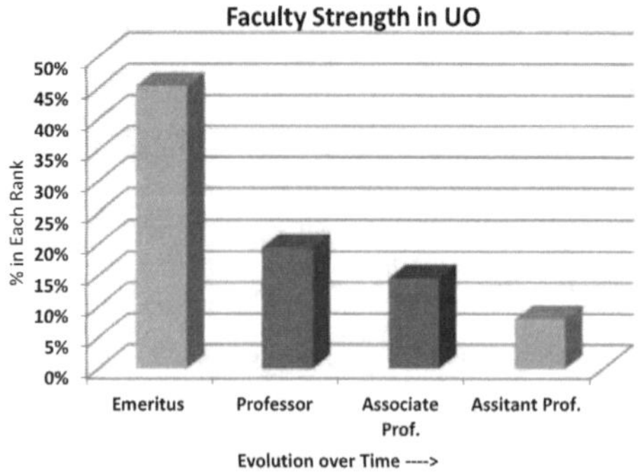

Aging demographic in particle technology and other unit operations

34

Survey on importance of skills (Rui Cruz – Dow)

Skill	Average Relative Importance 1-5
UO: unit operations, transport phenomena, thermodynamics, separation processes *	4.6
RE: reaction engineering, catalysis, kinetics.	4.0
AM: analysis, modeling, simulation, process control *	4.0
MAT: materials, surface science, polymers *	3.2
BIO: biotechnology, medical and life sciences	2.1
NANO: nanotechnology and its applications	1.8

** main perceived gaps between importance and proficiency by new hires*

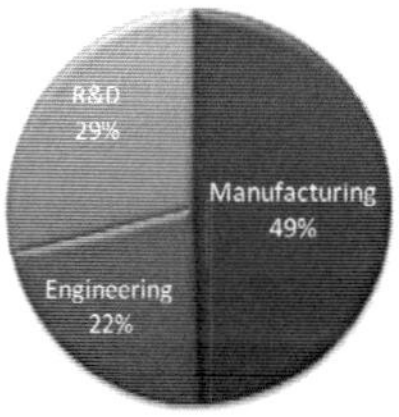

93 respondents (ChemE recruiters and leaders)

Dow

DOW RESTRICTED - For internal use only

35

Main concerns and opportunities

- Intimacy with equipment and mechanical skills
- ***Particle technology and solids handling***
- Applied statistics and experimental design
- Problem solving skills
- Communication, especially on technical matters
- Process safety, especially on reaction engineering
- Finance, process economics

- Coursework and project opportunities around application of chemical engineering fundamentals
- Foster learning and collaboration with other sciences and departments
- Industrial internships and co-op assignments

Dow

36

Final Thoughts

- Many great things have been accomplished in the last 25 years
- Still lots of work to do!
- Need continue the hard work in this important and fun area.

Acknowledgments

- Many thanks to my colleagues Madhu Kodam and Ben Freireich for some lively discussions around this topic. During these discussions, we reaffirmed that the sandbox was more than half full!
- Question asked by Ben's son, Rei, while playing in the sandbox at home, "Dad, do you like playing with dirt and sand?"

Controlling Particle Segregation:

Implications of Particle Cohesion, Properties, and Flow Perturbations

J. J. McCarthy

Department of Chemical Engineering
University of Pittsburgh

May, 2014

Acknowledgements

- Deliang Shi, Adetola Abatan, Watson Vargas, Suman Hajra, Tathagata Bhattacharya
- NSF: CBET-0933358, EEC-1005048
- ACS-Petroleum Research Fund, type AC
- DOEd, GAANN # P200A100087

Motivation

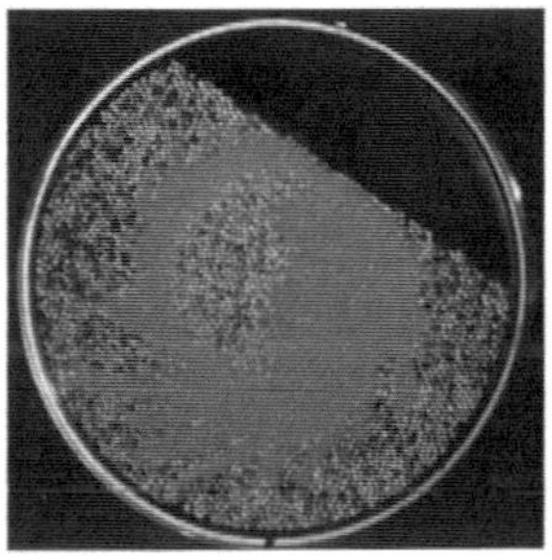

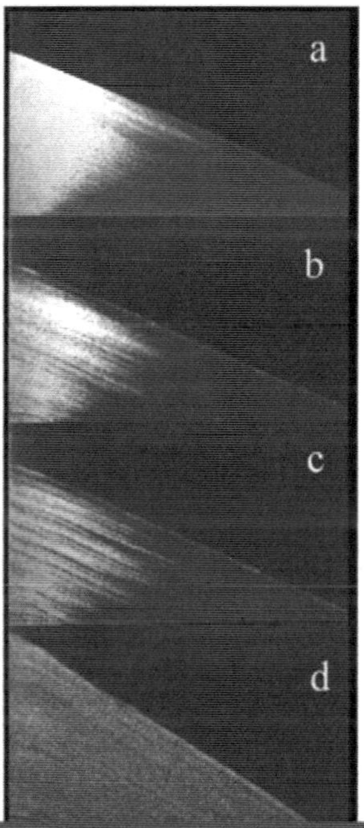

- Segregation ≡ un-mixing
- Industrial issues (product quality)
- Ubiquitous (even for "uniform" material)

Samadani and Kudrolli, *PRE* 2000

"Overmixing"

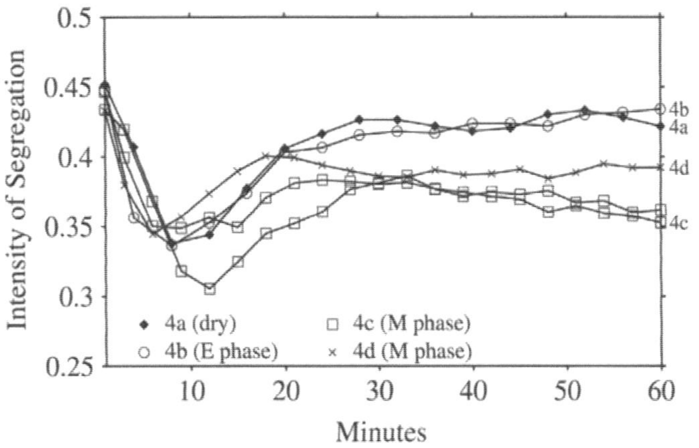

Li and McCarthy, *PT* 2006

- Segregation vs. mixing
- Rate/modeling is critical for scale up

Cohesion in Particle Systems

- **Surface and field forces**
  - van der Waals forces
  - Electrostatic forces
  - Magnetic forces
- **Material bridges**
  - liquid bridges
  - solid bridges
- **Interlocking by shape effects**

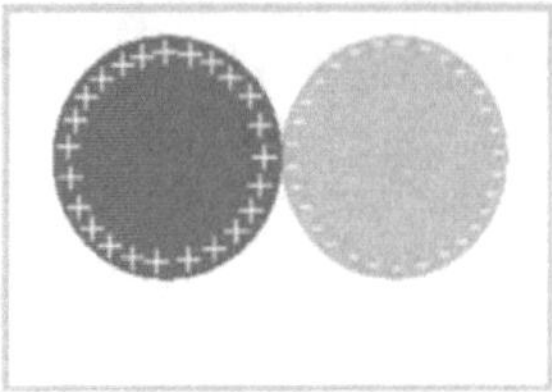

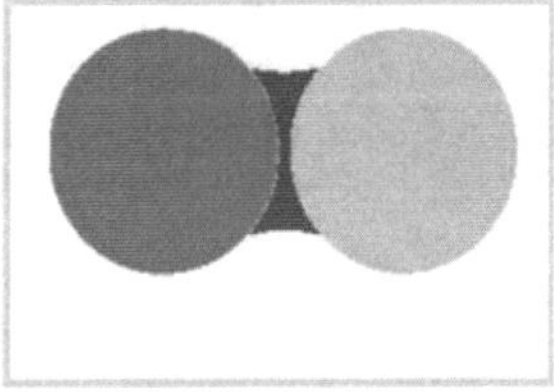

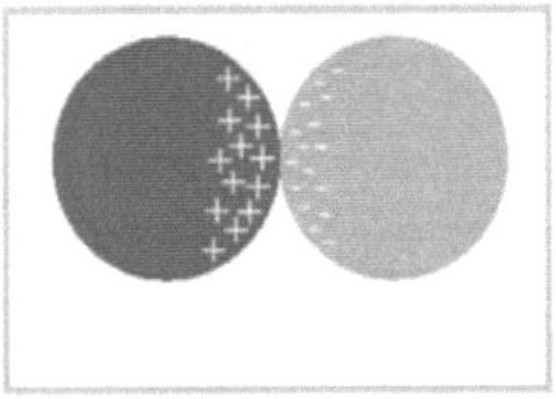

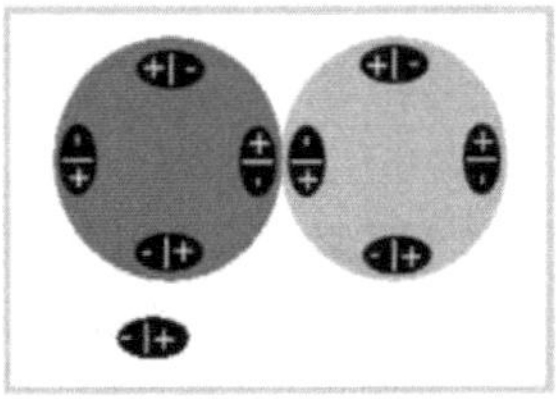

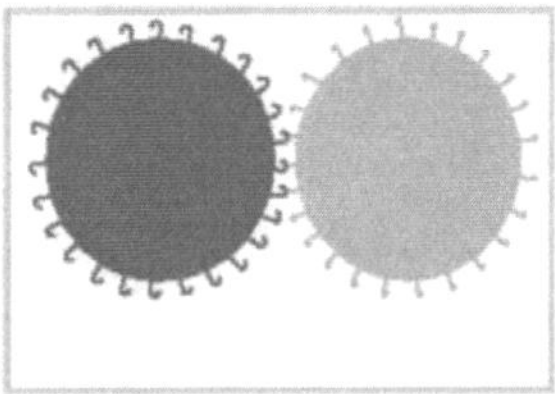

† Tomas J. Chem. Eng. Sci. Vol. 62, 7:1997-2010.

Quantifying Cohesion's Impact on Segregation – Bo_g

- **Granular Bond Number**

$$Bo_{ij} = \frac{F_{C_{ij}}}{F_W} = \frac{2R_{eff}\pi\Gamma_{ij}}{\frac{4}{3}\pi g[R^3\rho]_m} \text{ or } \frac{2\pi R_{eff}\gamma(cos\theta}{\frac{4}{3}\pi g[R^3\rho]_m}$$

Γ_{ij} : surface energy of particles
γ : surface tension of liquid
θ : contact angle
$(R^3\rho)_m$: the smaller of the two masses

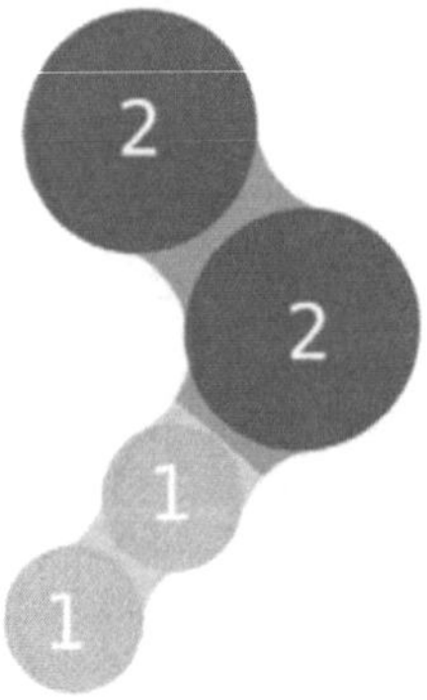

Li and McCarthy, *PRL*, 2003

Quantifying Cohesion's Impact on Segregation – Bo_g

- **Granular Bond Number**

$$Bo_{ij} = \frac{F_{C_{ij}}}{F_W} = \frac{2R_{eff}\pi\Gamma_{ij}}{\frac{4}{3}\pi g[R^3\rho]_m} \text{ or } \frac{2\pi R_{eff}\gamma(cos\theta}{\frac{4}{3}\pi g[R^3\rho]_n}$$

Γ_{ij} : surface energy of particles
γ : surface tension of liquid
θ : contact angle
$(R^3\rho)_m$: the smaller of the two masses

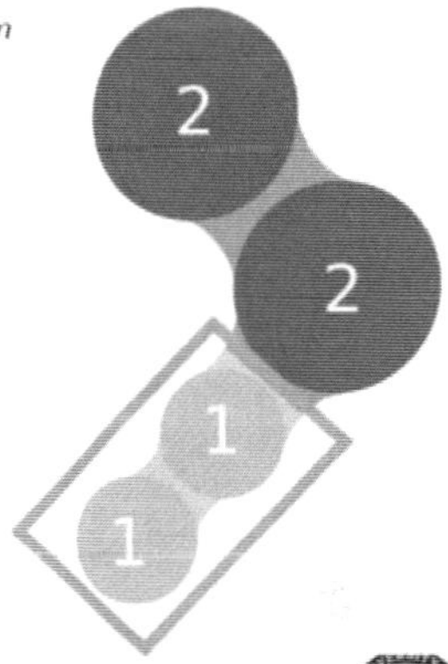

Li and McCarthy, *PRL*, 2003

Quantifying Cohesion's Impact on Segregation – Bo_g

- **Granular Bond Number**

$$Bo_{ij} = \frac{F_{C_{ij}}}{F_W} = \frac{2R_{eff}\pi\Gamma_{ij}}{\frac{4}{3}\pi g[R^3\rho]_m} \text{ or } \frac{2\pi R_{eff}\gamma(cos\theta)}{\frac{4}{3}\pi g[R^3\rho]_m}$$

Γ_{ij} : surface energy of particles
γ : surface tension of liquid
θ : contact angle
$(R^3\rho)_m$: the smaller of the two masses

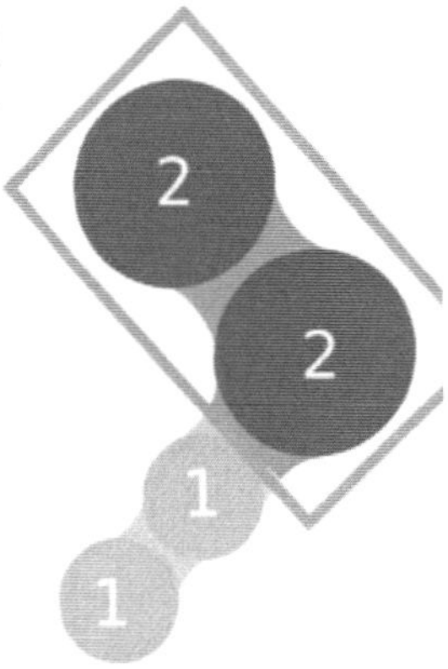

Li and McCarthy, *PRL*, 2003

Quantifying Cohesion's Impact on Segregation – Bo_g

- **Granular Bond Number**

$$Bo_{ij} = \frac{F_{C_{ij}}}{F_W} = \frac{2R_{eff}\pi\Gamma_{ij}}{\frac{4}{3}\pi g[R^3\rho]_m} \text{ or } \frac{2\pi R_{eff}\gamma(cos\theta)_i}{\frac{4}{3}\pi g[R^3\rho]_m}$$

Γ_{ij} : surface energy of particles
γ : surface tension of liquid
θ : contact angle
$(R^3\rho)_m$: the smaller of the two masses

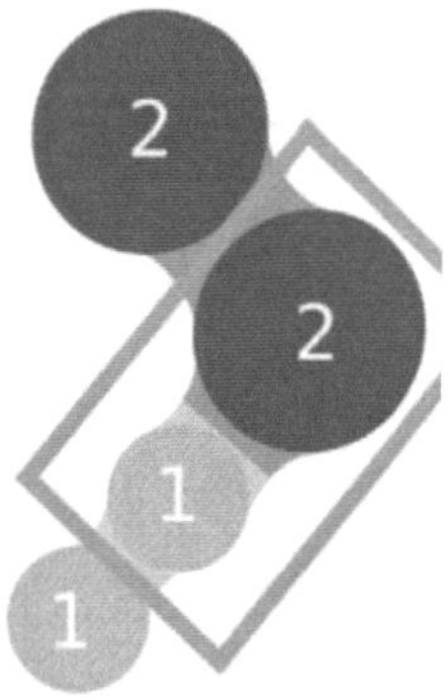

Li and McCarthy, *PRL*, 2003

Using This Info – Phase Diagrams

- **Mixed (M) phase**

$\Re_1 = \frac{Bo_{11}}{Bo_{12}} < 1$ and

$\Re_2 = \frac{Bo_{22}}{Bo_{12}} < 1$

- **Segregated (E) phase**

$\Re_1 = \frac{Bo_{11}}{Bo_{12}} > 1$ or

$\Re_2 = \frac{Bo_{22}}{Bo_{12}} > 1$

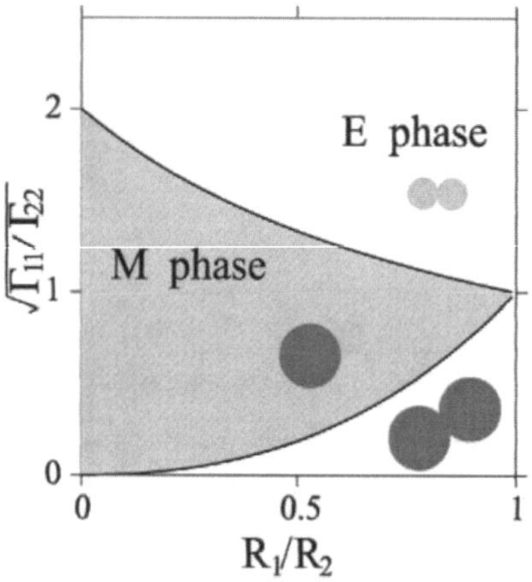

Assumptions:

- only considers isolated two-body interactions
- "thermalized system"
- all interactions are equally likely

Controlling Mix/Seg – Wet

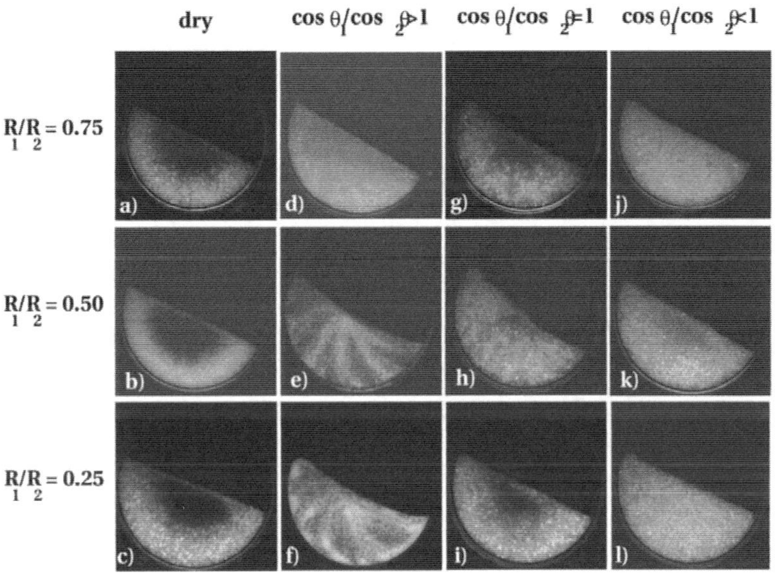

- Varying multiple particle and/or fluid properties

Motivation and Introduction
Asymptotic Behavior (Cohesive Segregation)
Segregation Rates (Free-flowing Materials)
Putting it all together (What's Next)
Modeling Cohesion
Controlling Segregation – Scaling Arguments
Controlling Segregation – Wet Systems

Controlling Mix/Seg – Wet

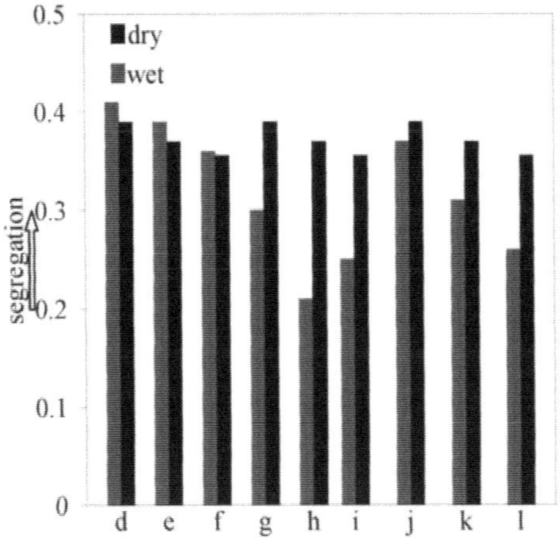

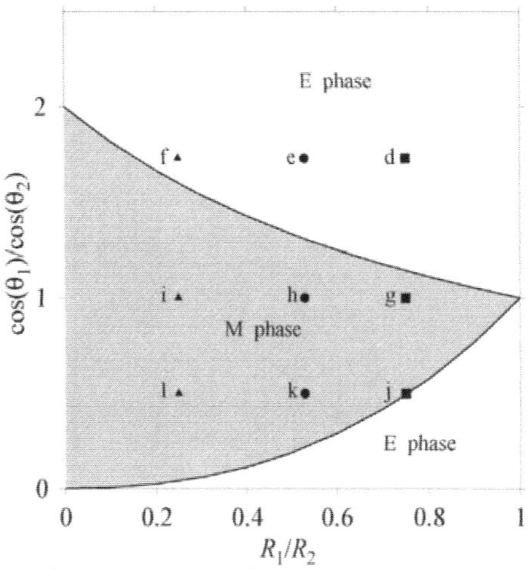

- "Control" by changing binder fluid (surfactants)
- Works also for wet "sheared" systems (use Co_g)

Li and McCarthy, *PRE*, 2005; Li and McCarthy, *PRE*, 2006

J. J. McCarthy Controlling Particle Segregation:

Controlling Mix/Seg – Wet

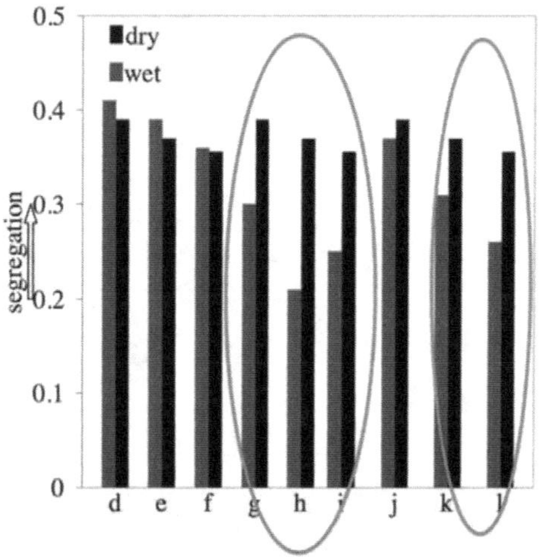

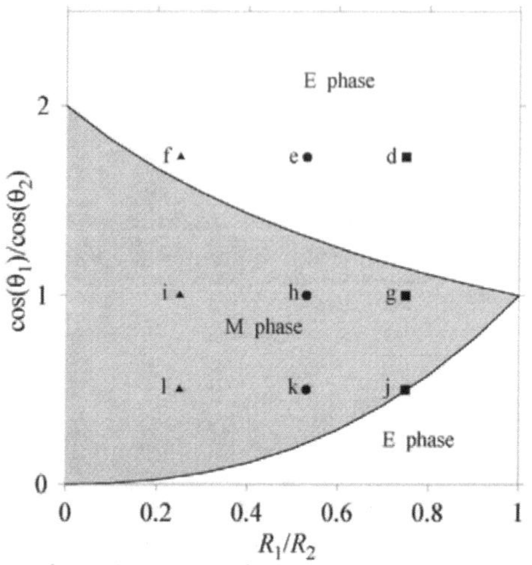

- "Control" by changing binder fluid (surfactants)
- Works also for wet "sheared" systems (use Co_g)

Li and McCarthy, *PRE*, 2005; Li and McCarthy, *PRE*, 2006

Free-Surface Segregation Characteristics

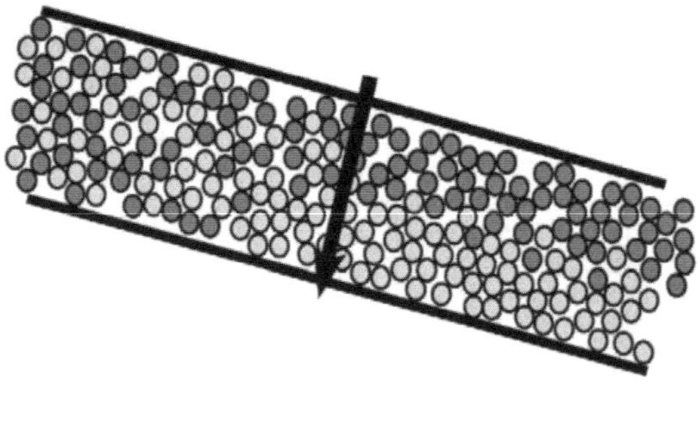

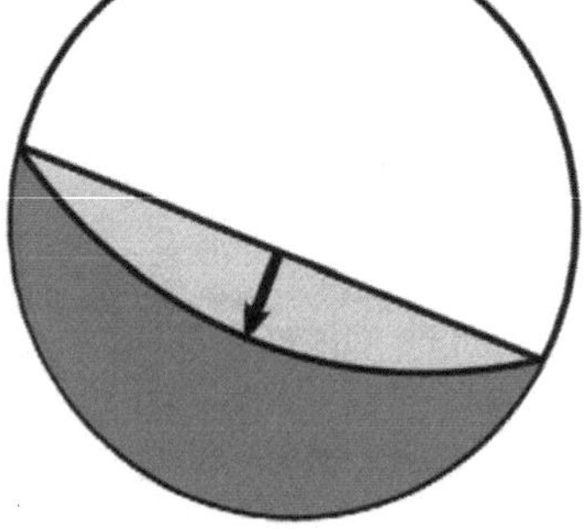

- segregation has a *preferred direction*
- segregation takes a finite time (t_s)
- Must exploit both!
  - change the direction of segregation
  - faster than t_s

Free-Surface Segregation Characteristics

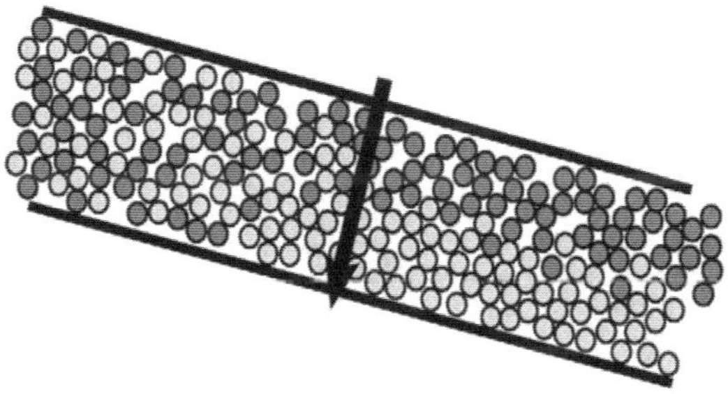

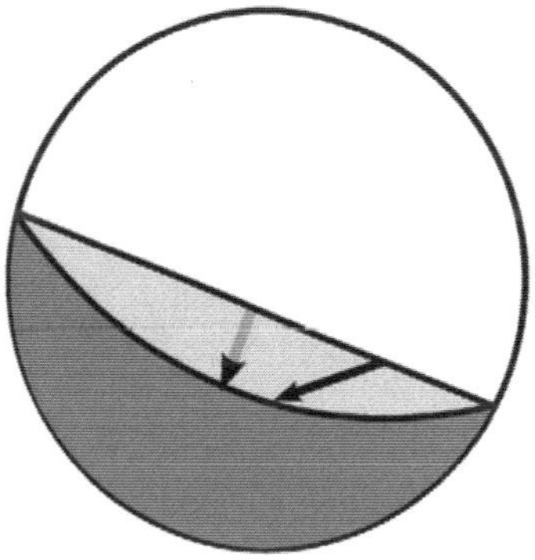

- segregation has a *preferred direction*
- segregation takes a finite time (t_s)
- Must exploit both!
  - change the direction of segregation
  - faster than t_s

Motivation and Introduction
Asymptotic Behavior (Cohesive Segregation)
Segregation Rates (Free-flowing Materials)
Putting it all together (What's Next)
Anti-segregation Arguments
Model Systems
A Practical System – A "Baffling" Tumbler
A Framework for Testing Segregation Theories

Free-Surface Segregation Characteristics

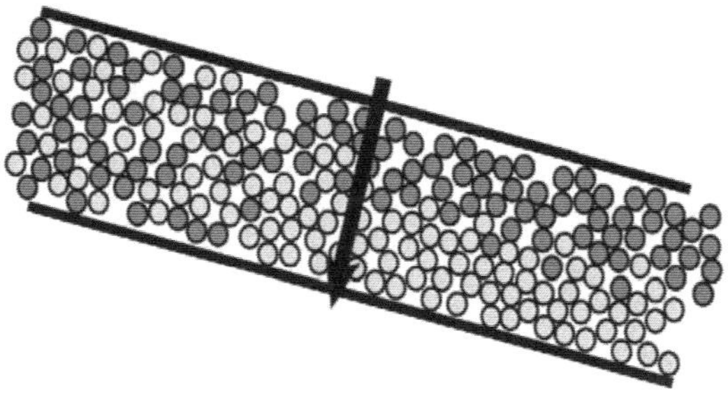

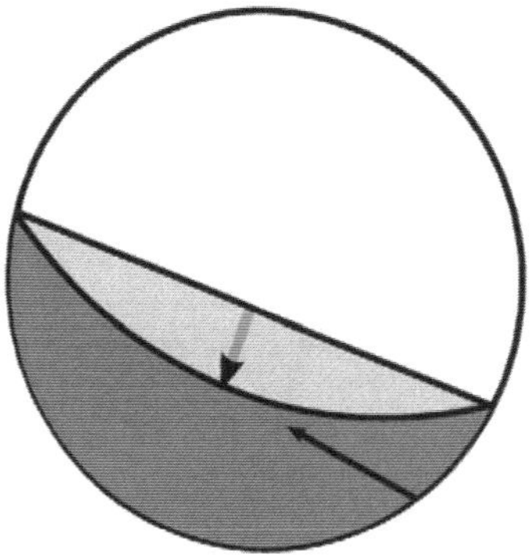

- segregation has a *preferred direction*
- segregation takes a finite time (t_s)
- Must exploit both!
  - change the direction of segregation
  - faster than t_s

Free-Surface Segregation Characteristics

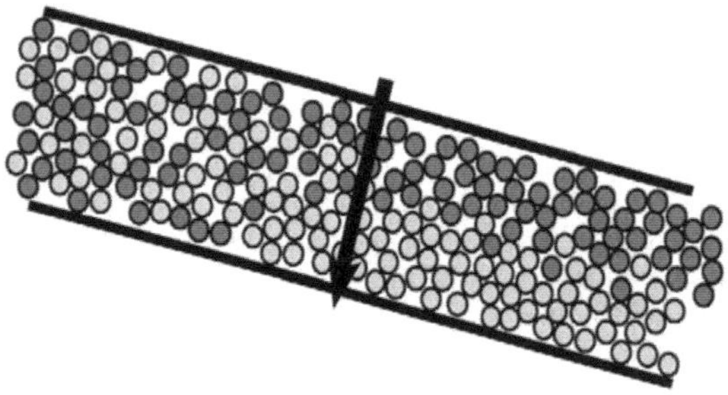

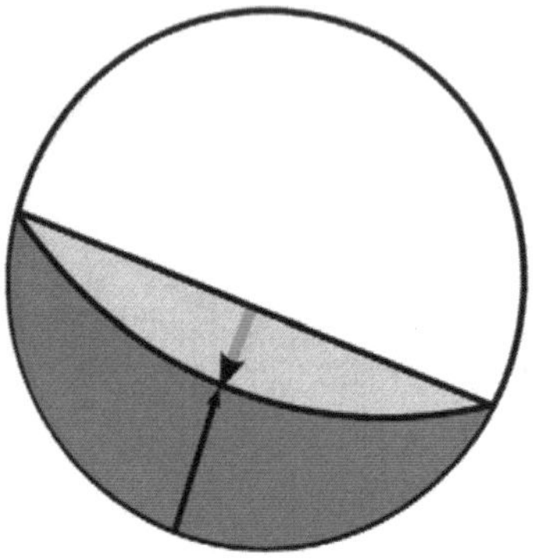

- segregation has a *preferred direction*
- segregation takes a finite time (t_s)
- Must exploit both!
  - change the direction of segregation
  - faster than t_s

Motivation and Introduction
Asymptotic Behavior (Cohesive Segregation)
Segregation Rates (Free-flowing Materials)
Putting it all together (What's Next)
Anti-segregation Arguments
Model Systems
A Practical System – A "Baffling" Tumbler
A Framework for Testing Segregation Theories

Free-Surface Segregation Characteristics

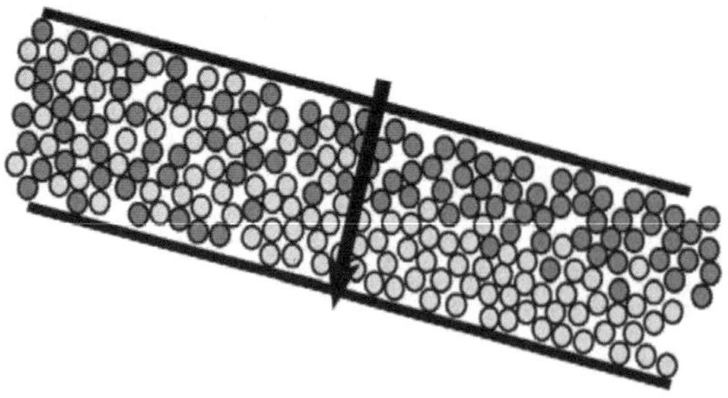

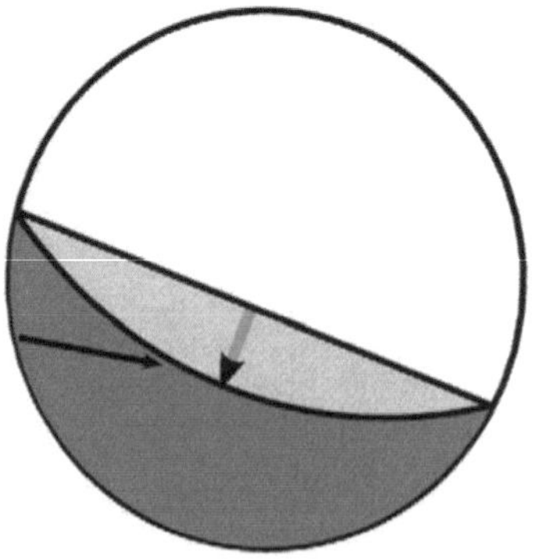

- segregation has a *preferred direction*
- segregation takes a finite time (t_s)
- Must exploit both!
  - change the direction of segregation
  - faster than t_s

Free-Surface Segregation Characteristics

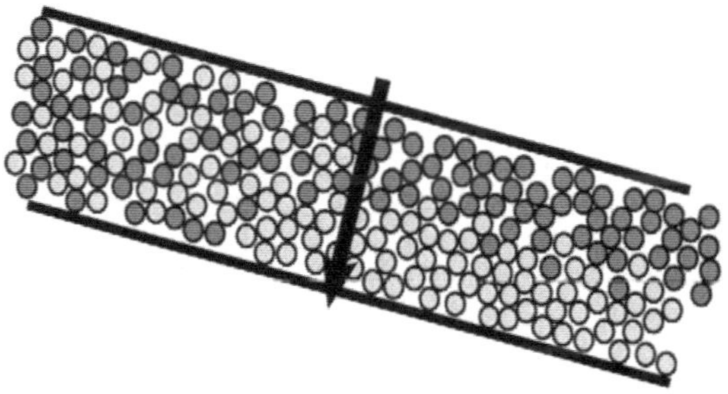

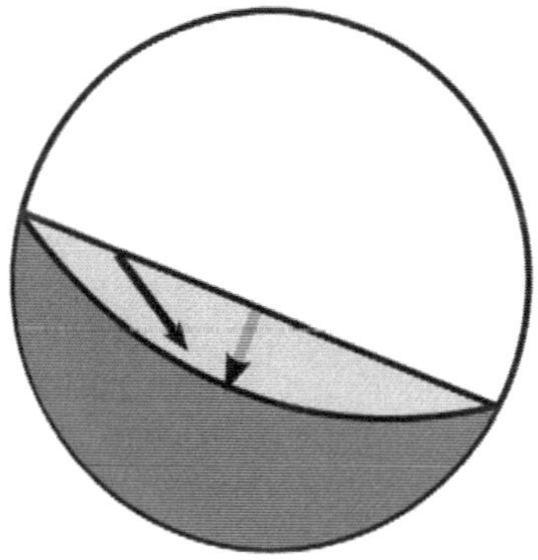

- segregation has a *preferred direction*
- segregation takes a finite time (t_s)
- Must exploit both!
  - change the direction of segregation
  - faster than t_s

"zig-zag" Chute Flow

- critical perturbation frequency $f_{crit} = t_s^{-1}$

  $f < f_{crit} \rightarrow$ segregating

  $f > f_{crit} \rightarrow$ mitigating segregation
- a "zig-zag" mixer

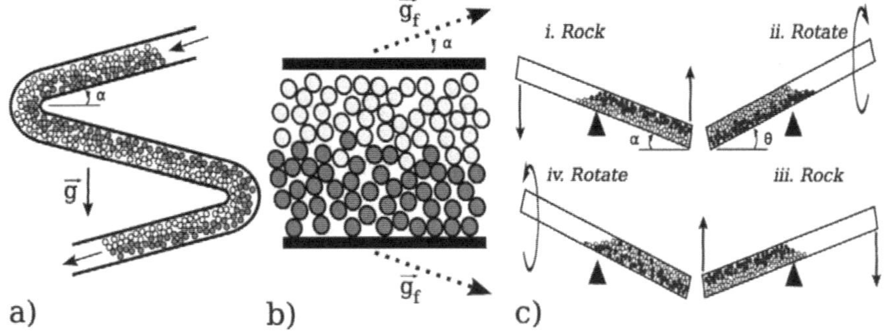

Shi, Abatan and McCarthy, *PRL*, 2007

Density Segregation Time – f_{crit}

- Segregation Péclet number
$$Pe = \frac{t_D}{t_s} = \frac{v_s R}{D}$$
where v_s is the segregation velocity, D is diffusivity
- Density segregation velocity
$$v_s = K_s(1 - \bar{\rho})$$
empirically $\frac{K_s R}{D}$ is O(1) so that $Pe \approx (1 - \bar{\rho})$
- Critical forcing frequency
$$f_{crit} = \frac{1}{t_s} = \frac{Pe}{t_D} \propto \frac{PeD}{R^2} \propto (1 - \bar{\rho})\dot{\gamma}$$
where $\dot{\gamma}$ is the shear rate and $D \propto R^2\dot{\gamma}$

Simulation Results – the zig-zag chute

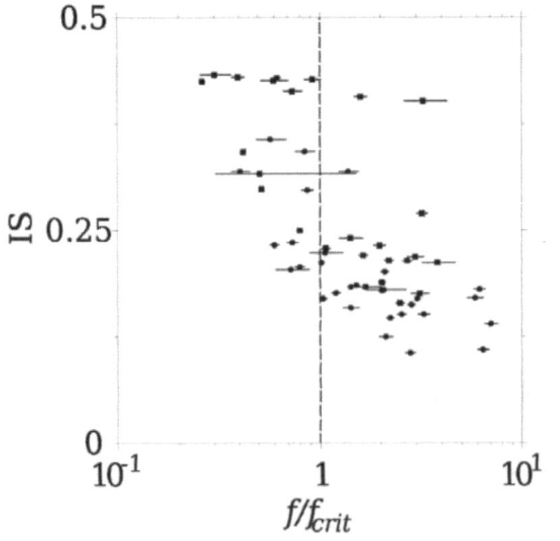

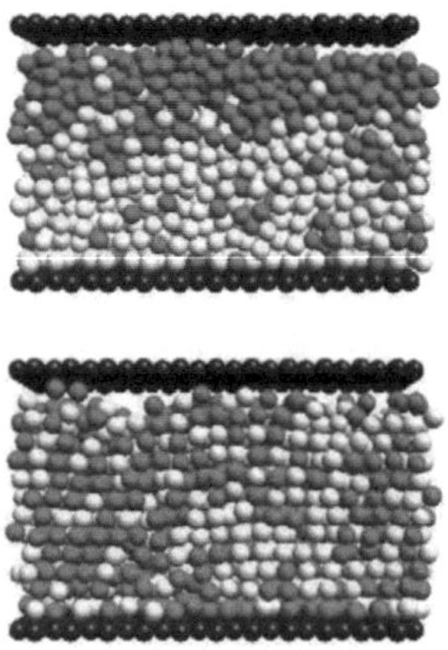

- Analyze more than 50 "geometries"
- Easily vary inclination angle, perturbation frequency, etc.

Baffling Results

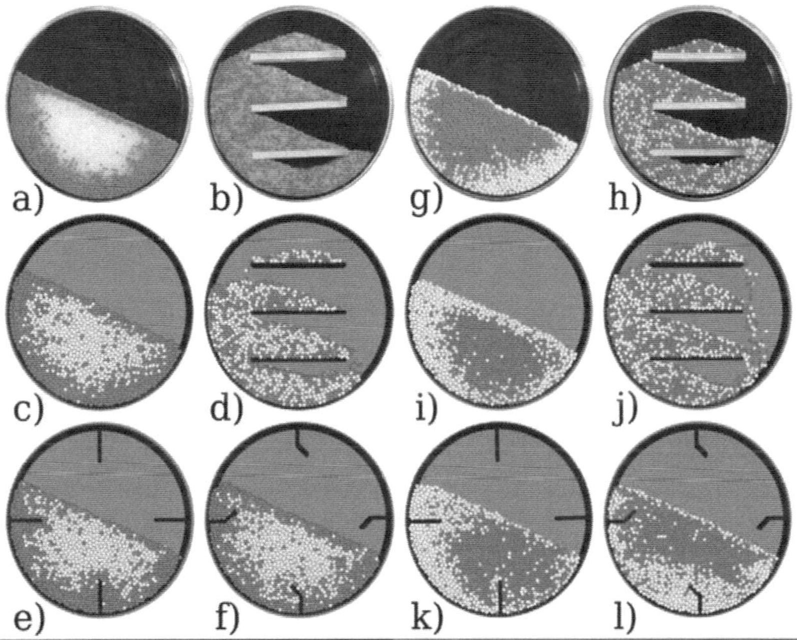

Tumbler – Density Segregation Theory

- Mean residence time

$$f = \frac{1}{\tau_{mean}} = \frac{\sqrt{\omega\dot{\gamma}}}{2\pi}$$

- Shear rate

$$\dot{\gamma} = \left[\frac{g\sin(\beta_m - \beta_s)}{cd\cos(\beta_s)}\right]^{1/2}$$

Khakhar and Ottino, 2002

- Recall, critical forcing frequency

$$f_{crit} = \frac{1}{t_s} \propto (1-\bar{\rho})\dot{\gamma}$$

- Frequency ratio

$$\frac{f}{f_{crit}} = \frac{K_1\sqrt{\omega}(d\cos_{\beta_S})^{1/4}}{(1-\bar{\rho})[g\sin(\beta_m - \beta_s)]^{1/4}}$$

$f < f_{crit} \rightarrow$ segregating

$f > f_{crit} \rightarrow$ mitigating segregation

Bhattacharya and McCarthy, 2014

Tumbler – Size Segregation Theory

- Size segregation velocity

$$v_s = [K_S + (1-\phi)K_T](1-\bar{d})$$

for fixed total concentration, ϕ, $v_s = [K_\phi](1-\bar{d})$

- Critical forcing frequency

by earlier logic, $f_{crit} \propto (1-\bar{d})\dot{\gamma}$

- Frequency ratio

$$\frac{f}{f_{crit}} \propto \frac{\sqrt{\omega}}{\sqrt{\dot{\gamma}}(1-\bar{d})}$$

$$\frac{f}{f_{crit}} = \frac{K_2\sqrt{\omega}(d_1 \cos_{\beta_S})^{1/4}}{(1-\bar{d})[g\sin(\beta_m - \beta_s)]^{1/4}}$$

Hajra, Bhattacharya and McCarthy, *Powder Tech.*, 2012

Effect of Density and Size Ratio

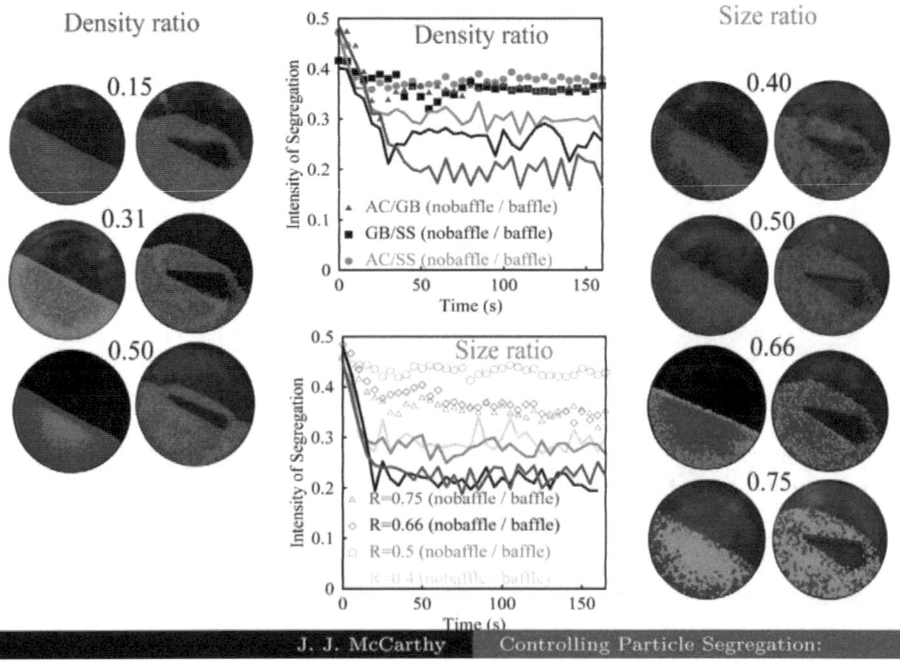

Effect of Cylinder Speed of Rotation

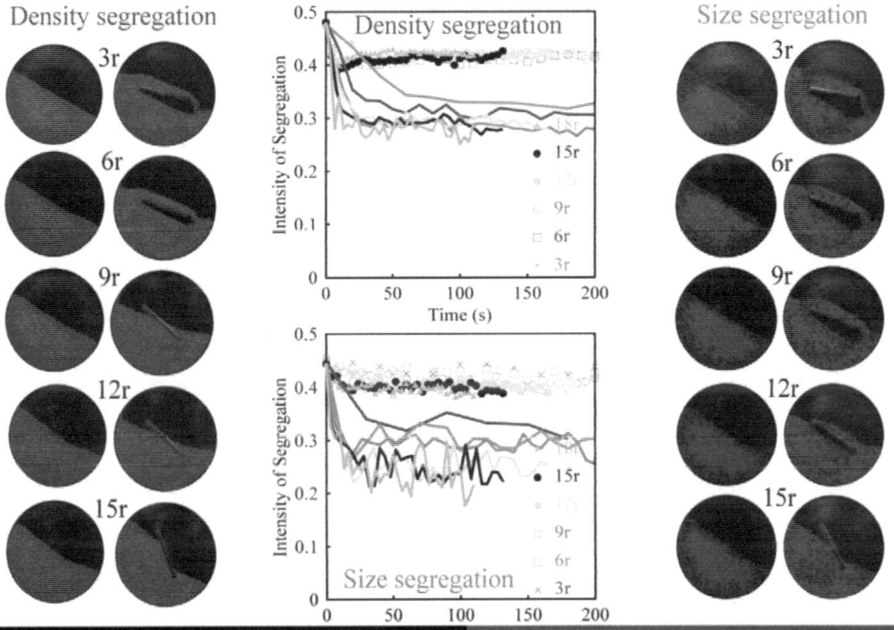

Model Prediction

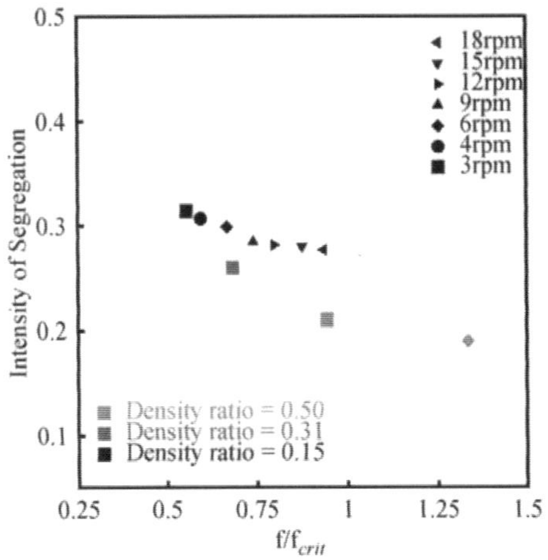

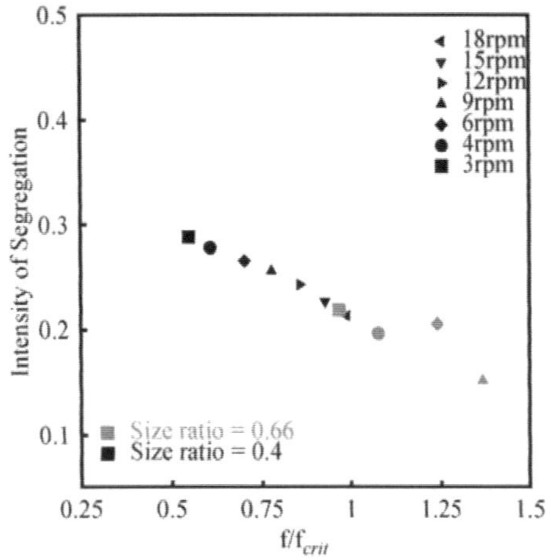

- Size model seems adequate
- Density model is a bit off

A New Density Segregation Model (Free-flowing)

- Force balance on particle

$$0 = F_w - F_b + F_d$$

where F_d is the drag force through bed (new!)

- Taking a single heavy (H) particle in a "sea" of light (L)

$$0 = m_H g \cos\theta - m_L g \cos\theta + c\pi\eta d_p v_{seg}$$

$\cos\theta$ is for inclination angle; η is "viscosity"; c is constant

- Solving for v_{seg}

$$\eta v_{seg} = -\frac{g\cos\theta}{c\pi d_p}(m_H - m_L)$$

so ηv_{seg} will be linear with $(m_H - m_L)$

Tripathi and Khakhar, *JFM* 2013

New Density Model (cont.)

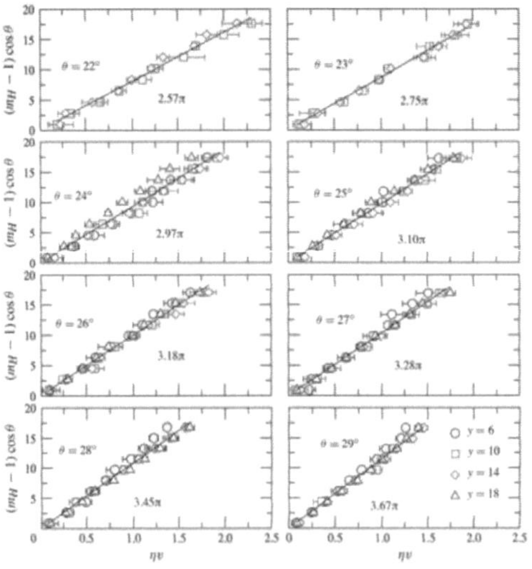

- Works for any inclination angle and density combo
- Needs to be scaled by "viscosity"

A **Proposed** Cohesive Density Segregation Model

- Force balance on particle **with cohesion**

$$0 = F_w - F_b - F_c + F_d$$

where F_c is the total cohesive force (new!)

- Taking a single heavy (H) particle in a "sea" of light (L)

$$0 = m_H g - m_L g - zF_c + c\pi\eta d_p v_{seg}$$

where z is the average coordination number (number of cohesive interactions)

- Solving for v_{seg}

$$v_{seg} = -\frac{W_L}{c\pi d_p \eta}\frac{(W_H - W_L - zF_c)}{W_L}$$

so v_{seg} will be linear with $\frac{(W_H - W_L - zF_c)}{W_L}$

Proposed Segregation Model (cont.)

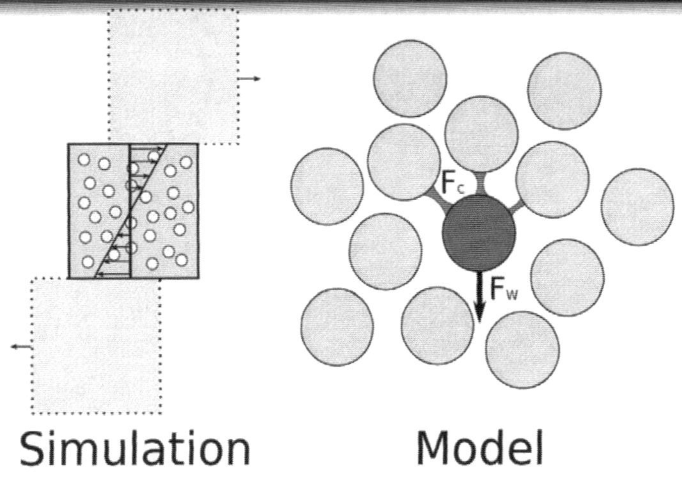

- Lees-Edwards bc eliminate need to "viscosity" (stays constant)
- Liquid bridges chosen as model cohesion

Testing Cohesive Density Rate Model (computationally)

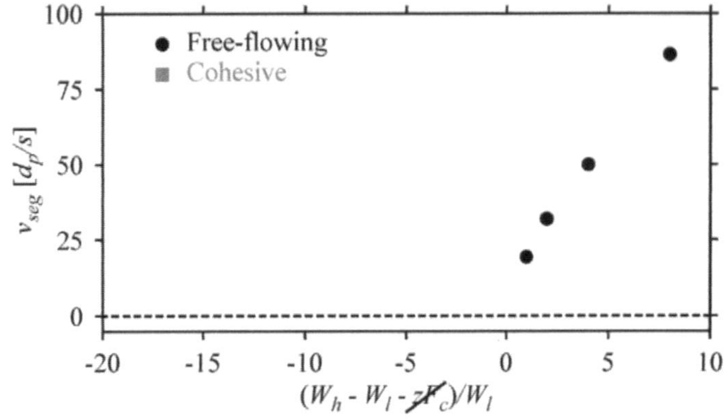

- Proposed model matches "new" model well
- Neither have been tested experimentally
- Segregation control framework should make this easy!

Motivation and Introduction
Asymptotic Behavior (Cohesive Segregation)
Segregation Rates (Free-flowing Materials)
Putting it all together (What's Next)

Testing Cohesive Density Rate Model (computationally)

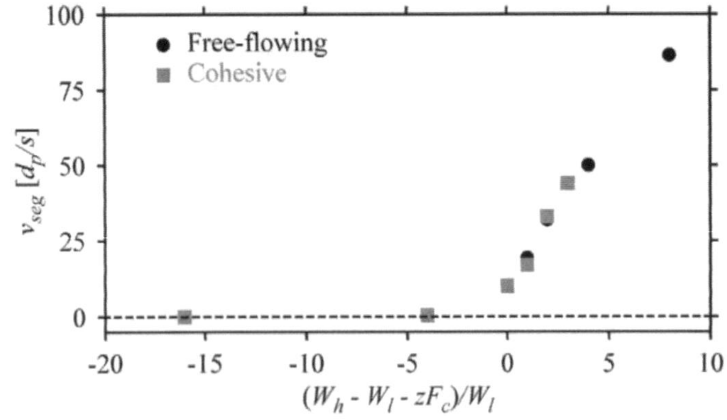

- Proposed model matches "new" model well
- Neither have been tested experimentally
- Segregation control framework should make this easy!

J. J. McCarthy Controlling Particle Segregation:

Outlook

- Cohesion Impacts Segregation Extent (Asymptotic Behavior)
  - $Bo_{12} > Bo_{ii} \rightarrow$ segregation extent is reduced
  - $Bo_{12} < Bo_{ii} \rightarrow$ segregation extent is increased!
- Segregation rate allows t_s vs "mixing" t_{mix}
  - $t_s > t_{mix} \rightarrow$ segregation doesn't occur
  - $t_s \approx t_{mix} \rightarrow$ "overmixing"
  - $t_s < t_{mix} \rightarrow$ strong segregation
- Segregation "patterns" (or lack thereof) is **obvious**
  - Segregation control framework $\rightarrow$ sensitive test of segregation kinetic models!
  - Allows simple experimental tests
  - Impact of cohesion on segregation **rate** is promising! (tested next)

Experimental Results – the rock-n-rotate mixer

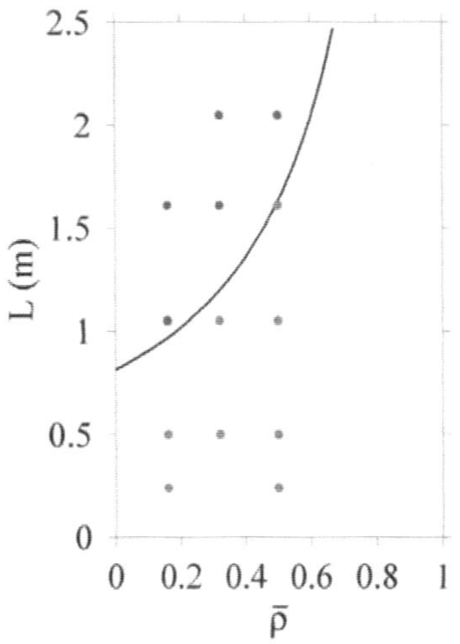

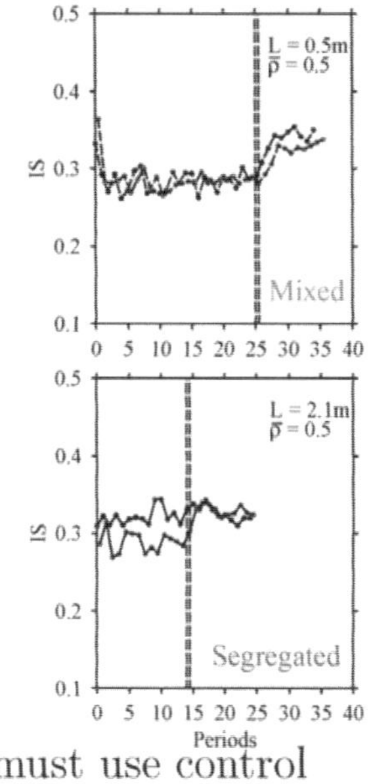

- $f/f_{crit} \propto L$
- Due to different mixing rates, must use control

Hajra, Shi, and McCarthy, *PRE*, 2012

Baffling Results (Quant)

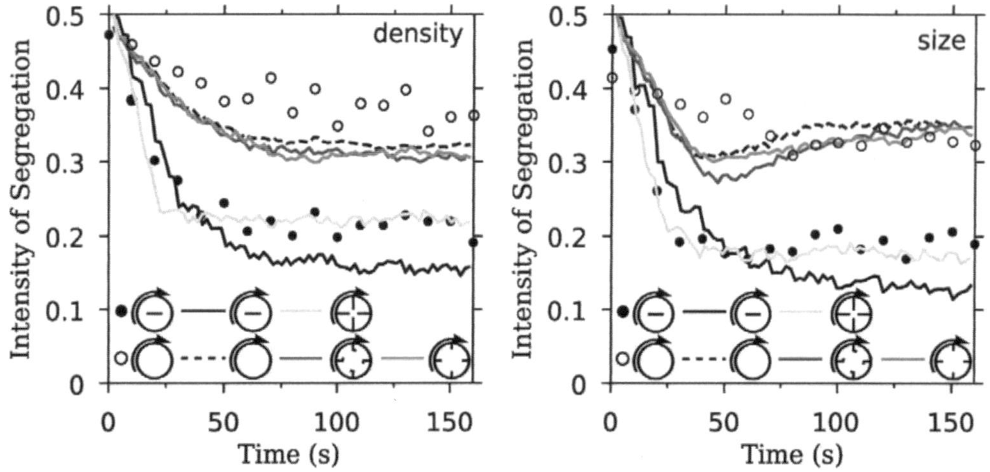

Vargas *et al.*, *AIChEJ*, 2008

Motivation and Introduction
Asymptotic Behavior (Cohesive Segregation)
Segregation Rates (Free-flowing Materials)
Putting it all together (What's Next)

Effect of Cylinder Fill Fraction

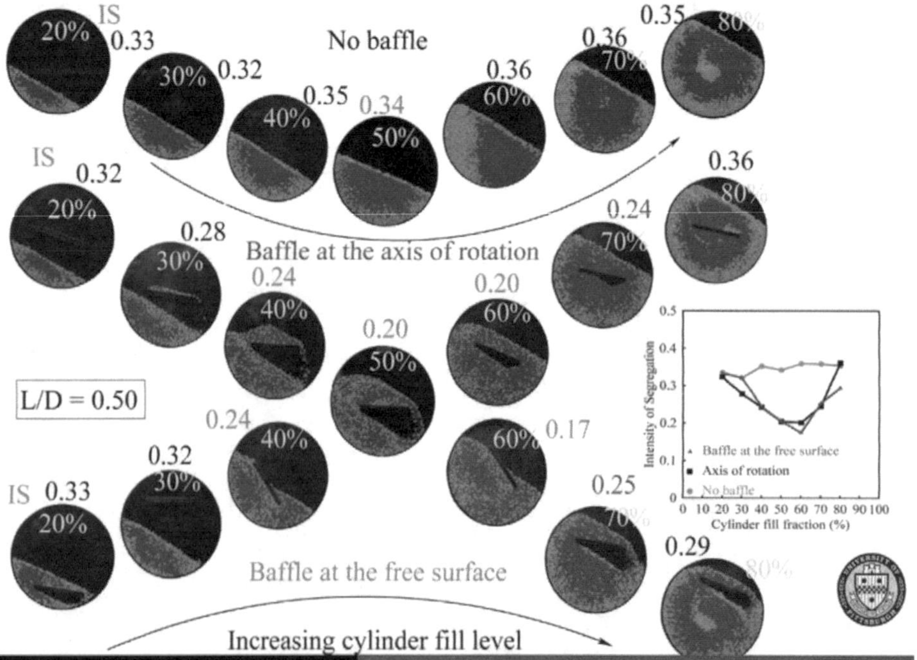

Shear Flow of Wet, Flexible Fibers using the Discrete Element Method

Jennifer Sinclair Curtis and Yu Guo
Department of Chemical Engineering, University of Florida

Carl Wassgren
School of Mechanical Engineering
Purdue University

Bruno Hancock and Bill Ketterhagen
Pfizer, Groton, CT

UF UNIVERSITY of FLORIDA
The Foundation for The Gator Nation

Hopper Discharge with Wet Spheres

Simulated granular flow from a 3-D hopper using DEM for spherical cohesive particles

- Explored the effects of **particle properties** like size and distribution, friction and coefficient of restitution
- Explored the effects of **hopper geometry** (hopper width, hopper angle, orifice size) and **hopper wall roughness**
- Characterized the **discharge rate** and **extent of segregation**

Quasi-3-D Plane Flow Hopper with Periodic Boundary Conditions

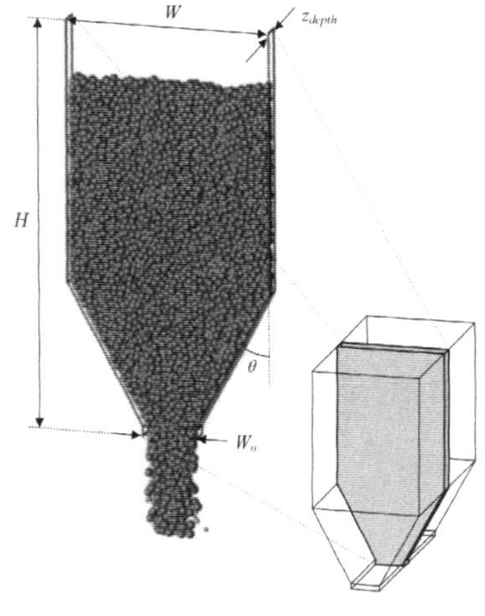

Anand, J. Curtis, C. Wassgren, B. Hancock, and W. Ketterhagen, "Predicting Discharge Dynamics of Wet Cohesive Particles from a Rectangular Hopper using the Discrete Element Method", *Chemical Engineering Science*, **64**, 5268-5275 (2009)

A. Anand, J. Curtis, C. Wassgren, B. Hancock, and W. Ketterhagen, "Segregation of Cohesive Granular Materials During Discharge from a Rectangular Hopper" (Invited), *Granular Matter*, **12**, 193-200 (2010)

2

DEM Model

Pendular Liquid Bridge Model

- Capillary force
  - Axial component of surface tension acting on the three-phase contact line
  - Hydrostatic force due to pressure deficiency in bridge
  - f (liquid bridge volume, particle separation)
    - Willett et al. (2000) equal-sized spheres
    - Soulie et al. (2006) two unequal-sized spheres
- Bridge rupture at a given separation distance
  - Varies linearly with contact angle and liquid bridge volume$^{1/3}$

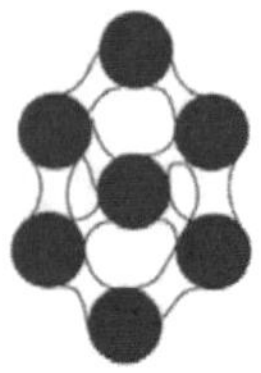

Soft-Particle Contact Model

- Normal Force-Displacement Model
  Hysteretic Spring – Walton & Braun[1]
- Tangential Force-Displacement Model
  Coulombic Friction & Rolling Resistance

[1] O. R. Walton & R. L. Braun, J. Rheol. **30**, 949 (1986).

3

Bond Number (Bo)

- Ratio of maximum cohesive force to the gravitational force
- Characterizes the cohesiveness of the system

$$Bo_g = \frac{|F_C|}{W} = \frac{2\pi a \gamma}{\frac{4}{3}\pi a^3 \rho g} = \frac{3\gamma}{2a^2 \rho g}$$

Water-Glass Bead System (a=2.2mm)
Bo=3.56

Bo ~ 10 for highly cohesive systems (McCarthy, 2003)

McCarthy J.J., Powder Technology, **138**, 63-67 (2003)

4

Segregation & Cohesion

Cohesionless (*Bo* = 0) Cohesion (*Bo* = 1)

- Depleted core of fines in the cohesionless case because of percolation, whereas, a more uniformly mixed state in the case of cohesion with no central depleted core of fines

6

Effect of Outlet Width and Bond Number

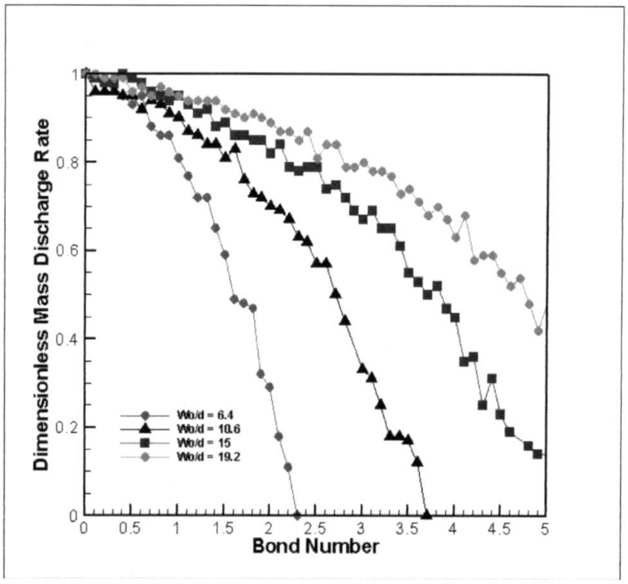

- Discharge rate decreases as the Bond number increases for a particular outlet width
- Deviation from cohesionless case is dependent on outlet width and Bond number

7

Modified Beverloo Correlation for Cohesive Particles

Discharge Rate for Rectangular Flat-Bottomed Hoppers - Cohesionless

$$\dot{W}_{flat-bottom} = 1.03\rho_{flow} g^{\frac{1}{2}}(L-kd)(W_O-kd)^{\frac{3}{2}}$$

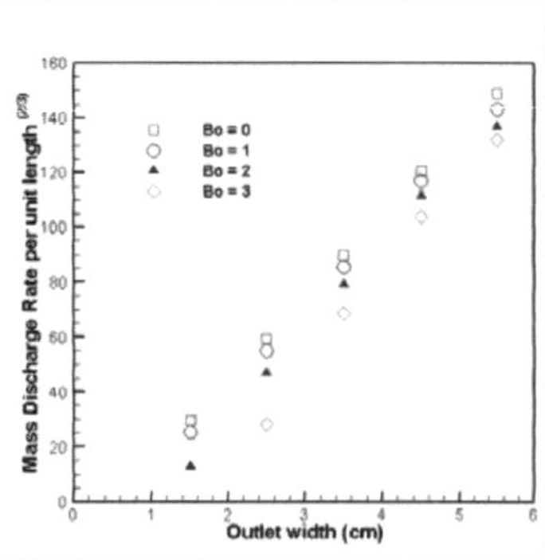

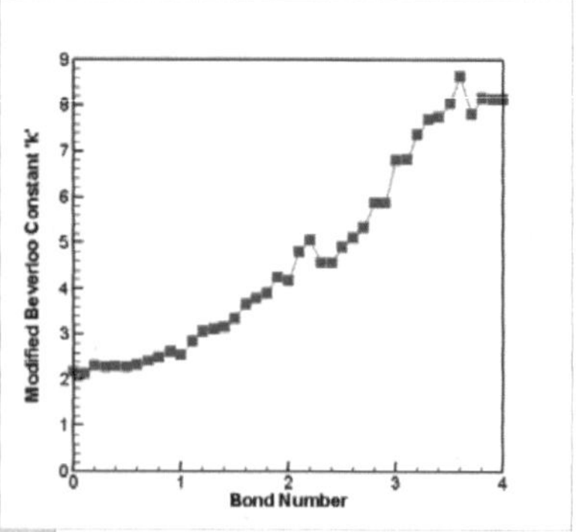

- Modified Beverloo constant increases with Bond number
- The apparent dead area near the outlet increases with cohesion because of formation of clumps

Discharge Rate for Rectangular Flat-Bottomed Hoppers - Cohesive

$$\dot{W}_{flat-bottom} = 1.03\rho_{flow} g^{\frac{1}{2}}[L_O-(1.9e^{0.39Bo})d][W_O-(1.9e^{0.39Bo})d]^{\frac{3}{2}}$$

Particle Shape and Cohesion

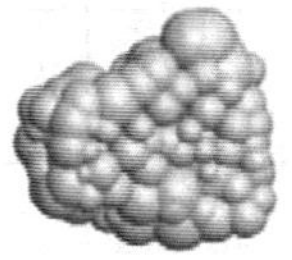

(1) Linked and Overlapping Spheres (Focused on Fibers)

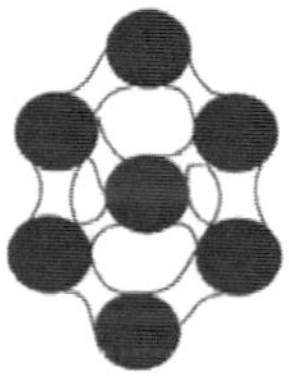

(3) Wet Surfaces

(2) Flexible Bonds between Constituent Particles

Includes all bond forces and moments – bending, stretching, twisting

UF | UNIVERSITY of FLORIDA
The Foundation for The Gator Nation

Flexible Fiber Model

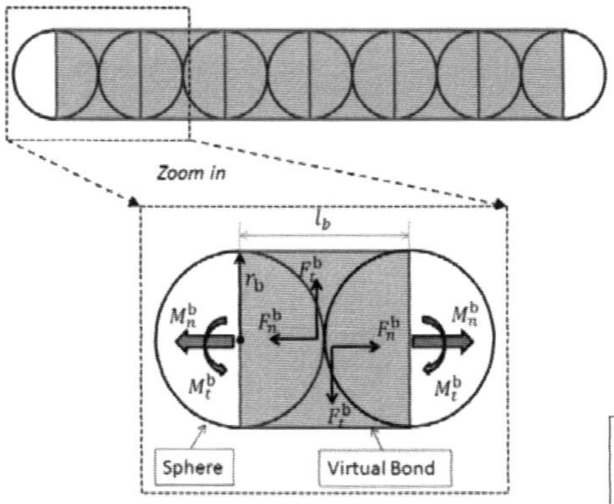

Bonded-sphere fiber model

Deforming flexible fiber

Normal Force Bending Moment	Shear Force Torsional Moment
$dF_n^b = \frac{E_b A}{l_b} d\delta_n^r$,	$dF_t^b = \frac{G_b A}{l_b} d\delta_t^r$,
$dM_t^b = \frac{E_b I}{l_b} d\theta_t^r$,	$dM_n^b = \frac{G_b I_p}{l_b} d\theta_n^r$.

Bonding forces/moments are computed incrementally (bonded-particle model of Potyondy and Cundall)

δ_n^r, δ_t^r	→ Relative normal and tangential displacements
θ_n^r, θ_t^r	→ Relative normal and tangential angular displacements
E_b, G_b	→ Bond Young's Modulus and Shear Modulus
A, l_b	→ Cross sectional area and length of a bond
I, I_p	→ Area moment of inertia for bending and polar area moment of inertia for twisting

Contact between two fibers is based on sphere-sphere contact (Hertz-Mindlin-Deresiewicz theory)

UF | UNIVERSITY of FLORIDA
The Foundation for The Gator Nation

Validation of Fiber Model

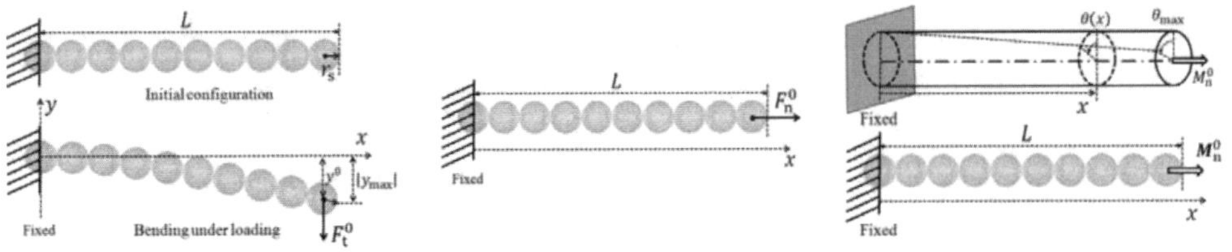

(a) Bending **(b) Axial stretching/compressing** **(c) Twisting**

- **Bending, axial stretching/compressing, and twisting deformations have been tested for a modeled fiber**
- **Static deformation and dynamic displacement have been examined**
- **Simulation results are in good agreement with analytical solutions, obtained from elastic beam theory**

Y. Guo, C. Wassgren, W. Ketterhagen, B. Hancock, and J. Curtis, "Validation and Time Step Determination of Discrete Element Modeling of Flexible Fibers," *Powder Technology*, **249**, 386-395 (2013)

UF | UNIVERSITY of FLORIDA
The Foundation for The Gator Nation

11

Particle Breakage

➢ **A bond breakage criterion can also be introduced:**

$$\sigma_{\max}^{b} = \frac{F_n^b}{A} + \frac{|M_t^b| r_b}{I} > \sigma_{ten}$$

OR $$\tau_{\max}^{b} = \frac{|F_t^b|}{A} + \frac{|M_n^b| r_b}{I_p} > \tau_{shear}$$

➢ **Bonding connection is eliminated when the maximum normal stress, σ_{max}^b exceeds the material tensile strength σ_{ten}**

➢ **Bonding connection is eliminated when the maximum shear stress τ_{max}^b exceeds the shear strength τ_{shear}**

[1] D.O., Potyondy, P.A., Cundall. *International Journal of Rock Mechanics & Mining Sciences* 41 (2004) 1329-1364.

UF | UNIVERSITY of FLORIDA
The Foundation for The Gator Nation

12

Compaction of Chalk Sticks

Experiment

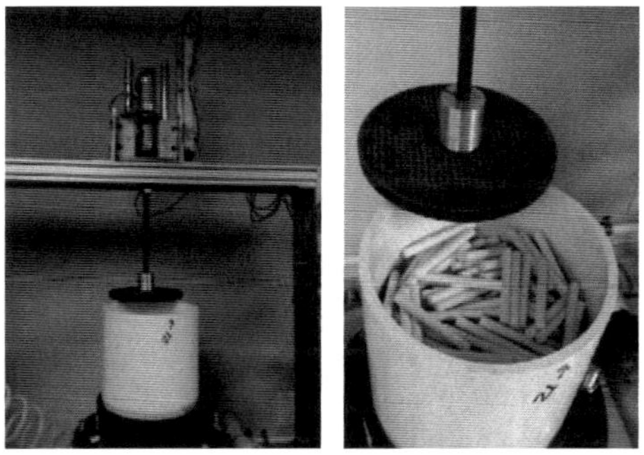

Simulation

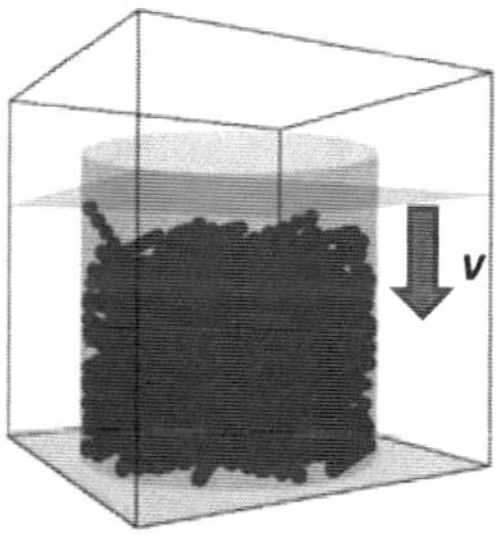

- Before compaction

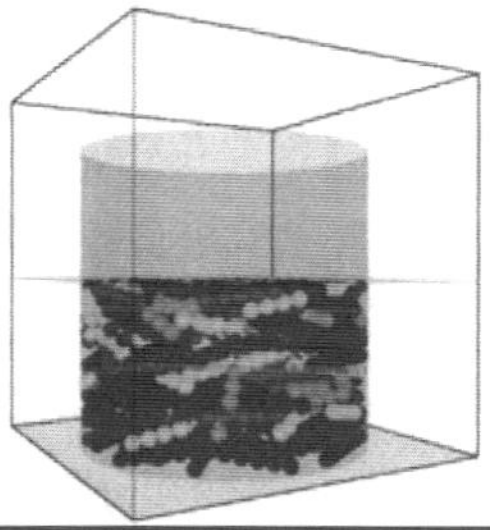

- During compaction

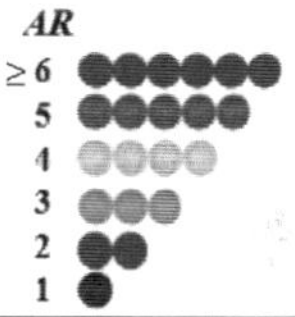

- Container dimensions: D=200 mm, H=195 mm
- Chalk dimensions: d = 10 mm, L = 80 mm
- # of chalks = 300
- Apply load at a rate of 5-10 mm/s

Experiments conducted in *Materials Flow Solutions, Inc.* (Gainesville, FL)

UF UNIVERSITY of FLORIDA
The Foundation for The Gator Nation

Comparison between Simulations and Experiments

Load *V.S.* Displacement of Punch

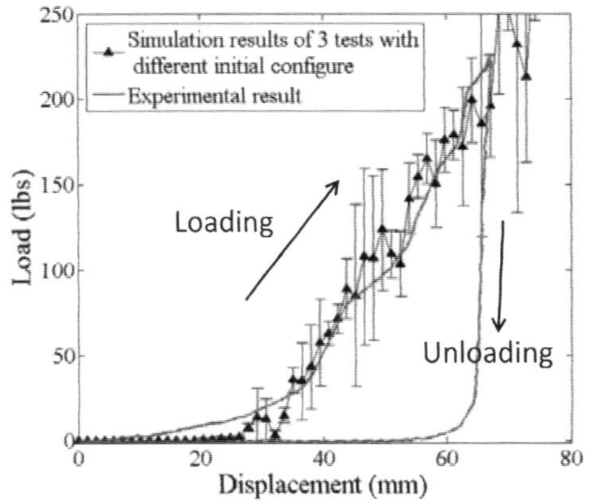

Particle Size Distribution at 230 lbs

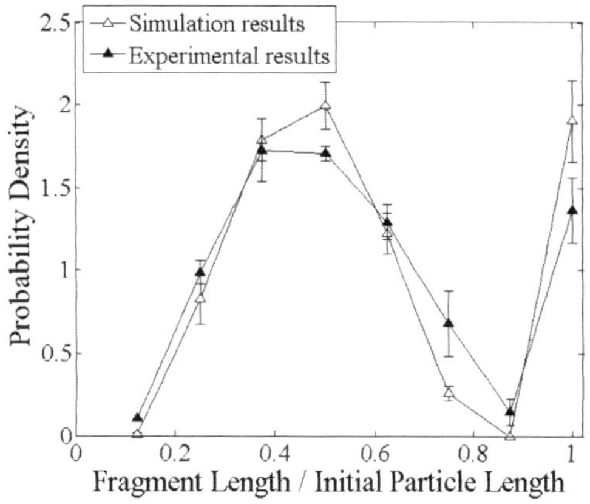

➢ **Good agreement between simulations and experiments**

UF UNIVERSITY of FLORIDA
The Foundation for The Gator Nation

Particle Breakage in Agitation Cell

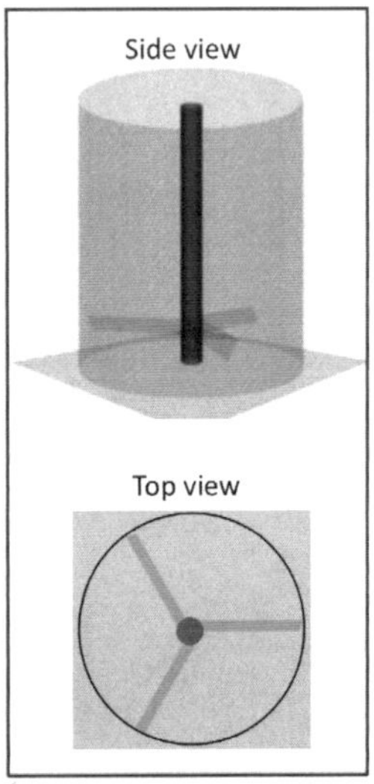

Cylindrical cell with an impeller

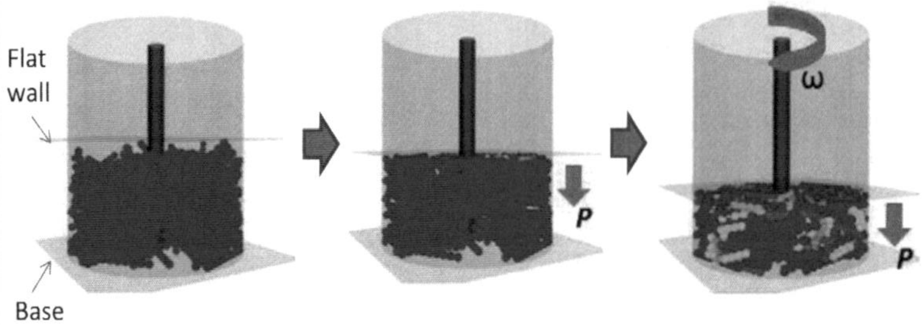

- A number of elongated particles with AR = 10 are generated in the cell and deposited under gravity
- Particle bed is compressed by applying a weight on the top
- The particle bed, which remains compressed, is agitated by rotating the impeller

UF | UNIVERSITY of FLORIDA
The Foundation for The Gator Nation

Particle Breakage by Agitation under Different Pressures

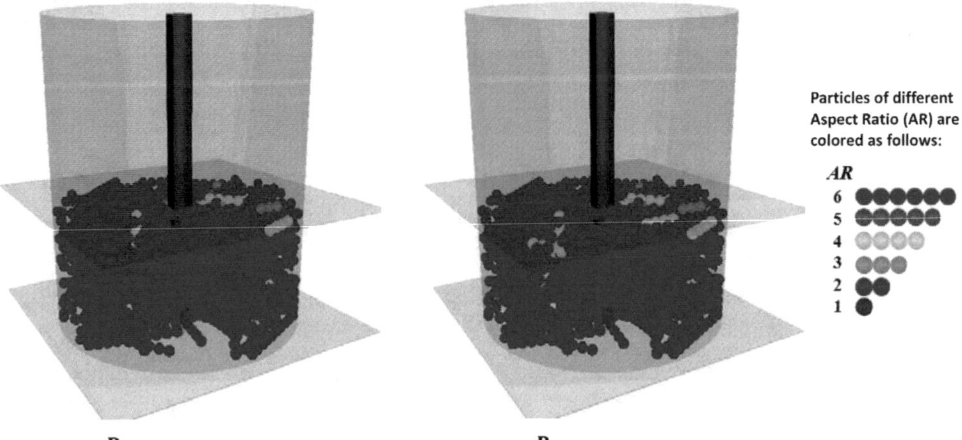

$\frac{P}{\sigma_{ten}} = 8.45\,e-4$ (Low pressure) $\frac{P}{\sigma_{ten}} = 1.69\,e-3$ (High pressure)

(Agitation cell diameter D=122 mm, $\mu_{pp} = 0.2$, $\mu_{wp} = 0.7$, ω = 60 rpm)

- **In a given number of impeller rotations, more particles break under higher pressure**

UF | UNIVERSITY of FLORIDA
The Foundation for The Gator Nation

Effect of Wall-Particle Friction

$\mu_{wp} = 0.7$ $\mu_{wp} = 0.2$

Particles of different Aspect Ratio (AR) are colored as follows:

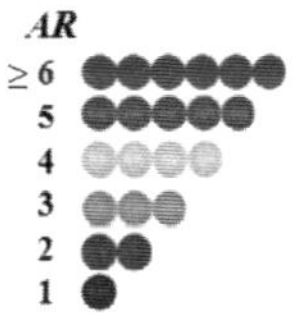

(Agitation cell diameter D=122 mm, $\mu_{pp} = 0.2$, ω = 60 rpm, $\frac{P}{\sigma_{ten}} = 1.69\,e - 3$)

- **With a small wall-particle friction, particle bed may rotate like a solid body with impeller, leading to rare occurrence of breakage**

UF | UNIVERSITY of FLORIDA
The Foundation for The Gator Nation

Comparison between Simulations and Experiments

Experimental results*

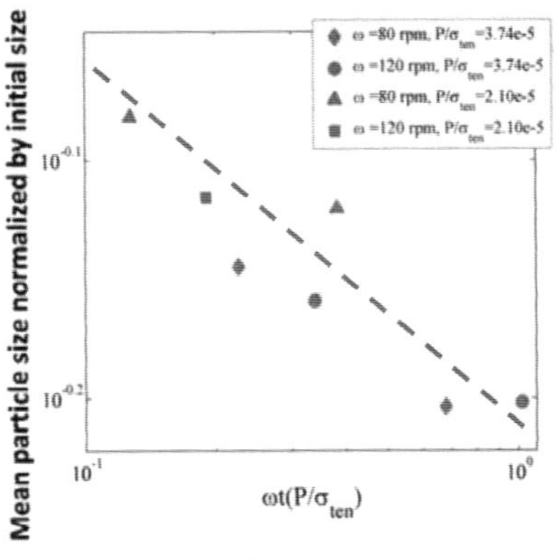

- Material: **L-threonine crystals**
- Initial mean particle size: **385 μm**
- Agitation cell diameter: **25 mm**

Simulation results

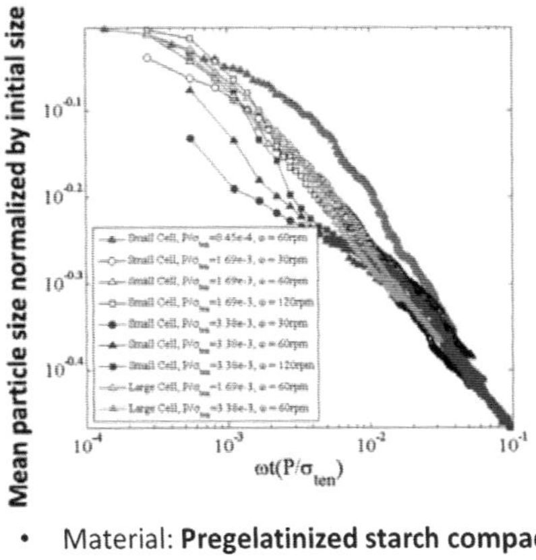

- Material: **Pregelatinized starch compacts**
- Initial mean particle size: **50 mm**
- Agitation cell diameter: **122 mm, 244mm**

➢Particle size reduction depends on cumulative work

** The agitation cell experimental data were provided by the Pfizer Inc.*

UF | UNIVERSITY of FLORIDA
The Foundation for The Gator Nation

Shear Flows of Flexible Fibers

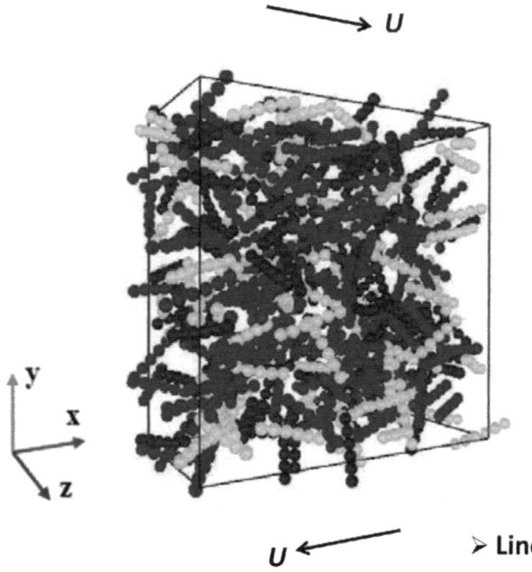

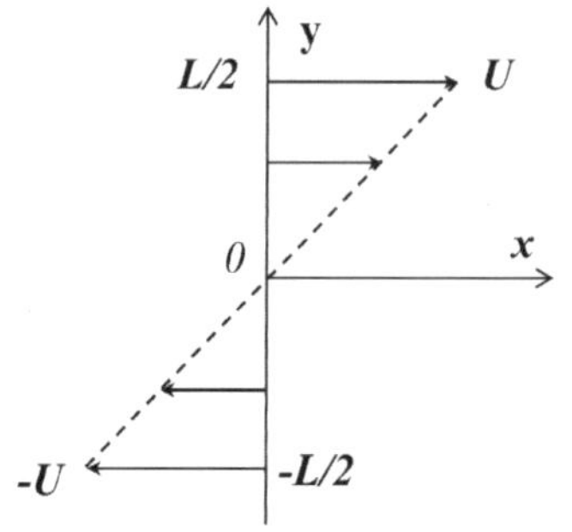

X-velocity vector distribution

- **Linear x-velocity profile varying in y direction**
- **Periodic boundary conditions in x and z directions**
- **Lees-Edwards boundary condition in y direction**

Fiber Deformation at Various Solid Fractions

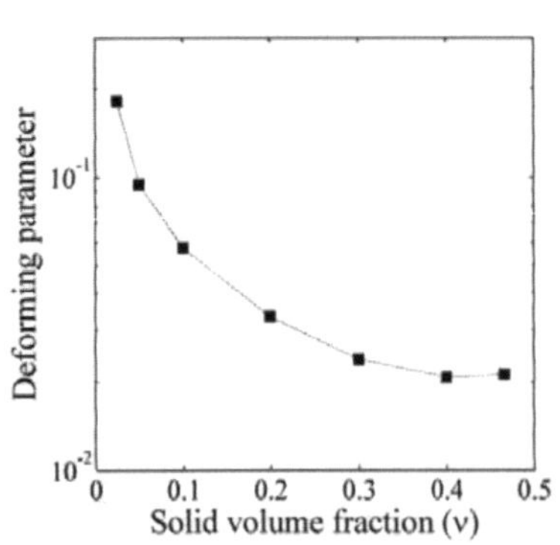

(AR = 6, E_b = 2.6e4 Pa, β_b = 0, β_c = 1.63e-2, μ = 0)

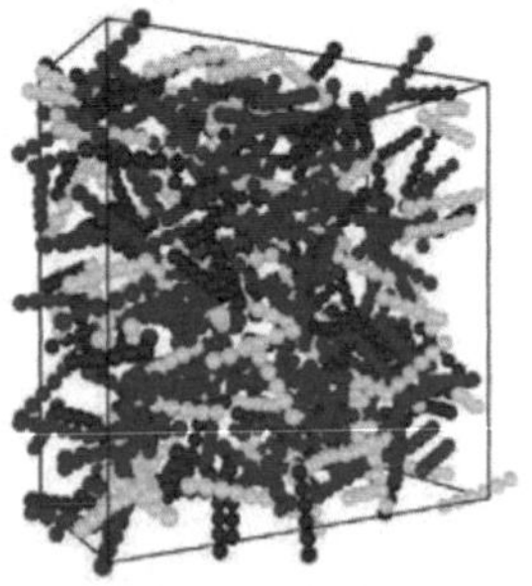

- Solid Fraction = 0.1
- Deforming parameter approaching 1, particularly as bond stiffness decreases

- Solid Fraction = 0.466
- Alignment of fibers in the flow direction occurs and deforming parameter becomes very small

Fiber-Phase Stress

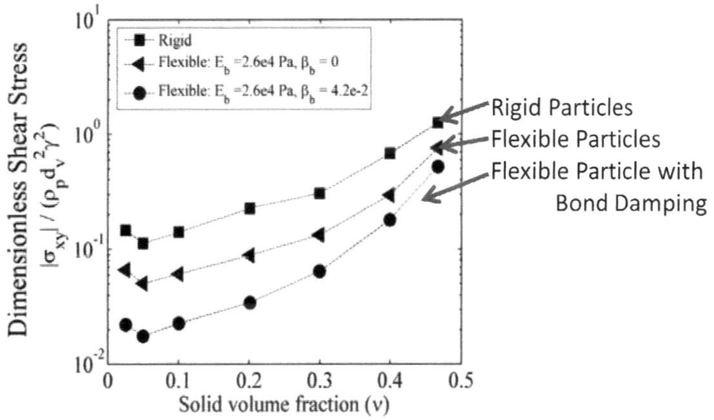

(AR = 6, E_b = 2.6e4 Pa, β_c = 1.63e-2, μ = 0)

- Similar trends are observed for the variation of stress with solid volume fraction with rigid and flexible fibers
- Smaller stresses are obtained for flexible fibers compared to rigid fibers
- Stresses decrease with bond damping (β_b)

UF | UNIVERSITY of FLORIDA
The Foundation for The Gator Nation

21

Effect of Particle Friction

Aligning and Layer Flow | **Jostling and Entangling Flow**

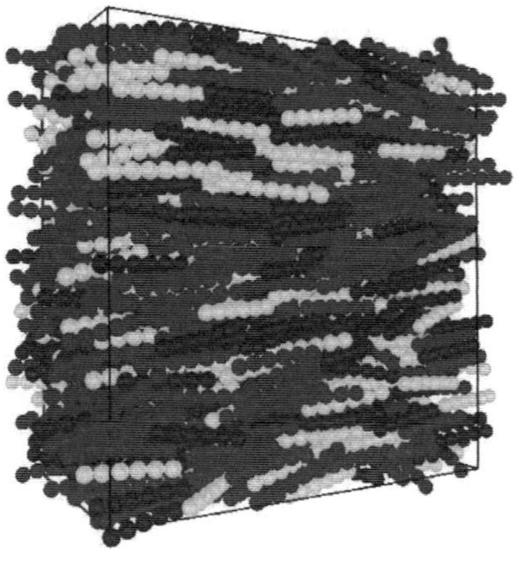

Without friction (μ = 0)

With friction (μ = 0.5)

(Fiber Aspect Ratio AR = 6, E_b = 2.6 e4 Pa, β_c = 1.63e-2, β_b = 0, Solid Fraction = 0.4663)

UF | UNIVERSITY of FLORIDA
The Foundation for The Gator Nation

22

Effect of Fiber Aspect Ratio

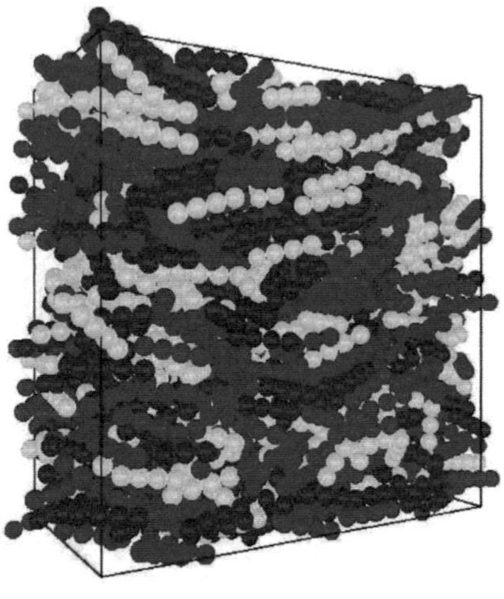

Fiber Aspect Ratio *AR* = 4

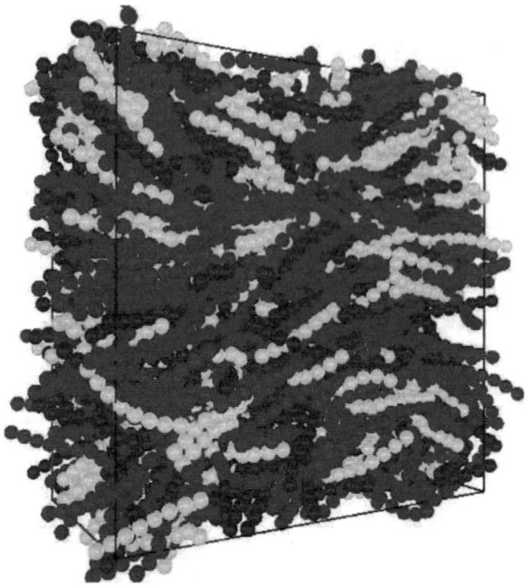

Fiber Aspect Ratio *AR* = 8

(Solid Volume Fraction = 0.4, E_b = 1.0e8 Pa, β_c = 1.63e-2, β_b = 0 , μ = 0.5)

UF UNIVERSITY of FLORIDA
The Foundation for The Gator Nation

23

Effect of Fiber Aspect Ratio

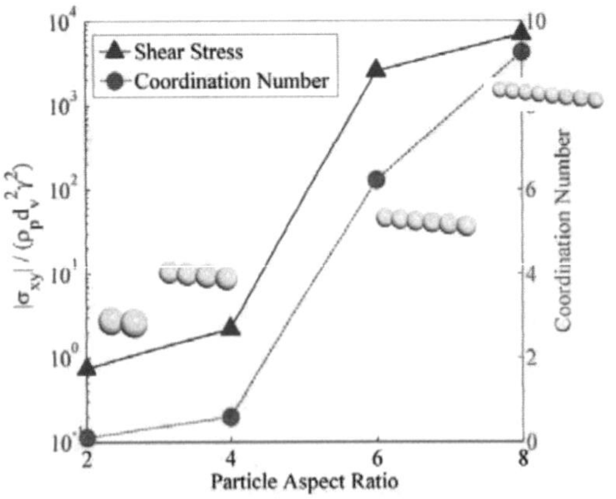

(Solid Volume Fraction = 0.4, E_b = 1.0e8 Pa, β_c = 1.63e-2, β_b = 0 , μ = 0.5)

In dense flows of flexible fibers, as particle aspect ratio increases,

- Shear Stress
- Coordination Number

↑

UF UNIVERSITY of FLORIDA
The Foundation for The Gator Nation

24

Effect of Particle Surface Roughness

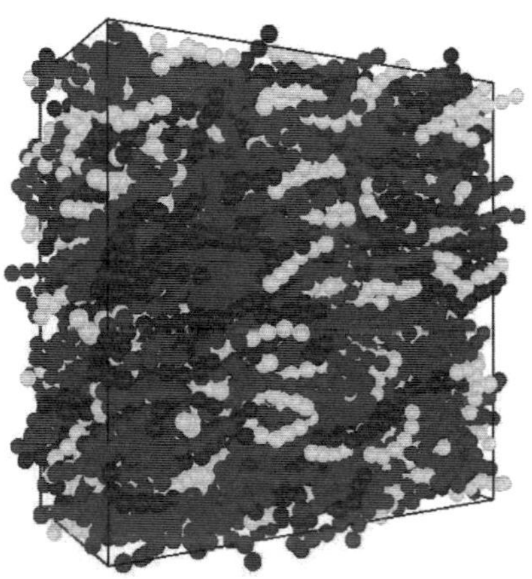

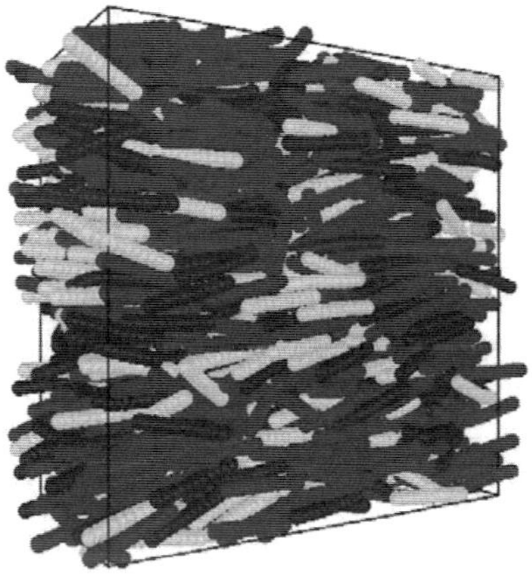

Entangling Flow with Rough Particles (6-sphere particle)

Layer Flow with Smooth Particles (11-sphere particle)

(Fiber Aspect Ratio AR = 6, E_b = 2.6 e4 Pa, β_c = 1.63e-2, β_b = 0, μ = 0.5, Solid Fraction = 0.4663)

UF UNIVERSITY of FLORIDA
The Foundation for The Gator Nation

25

Effect of Particle Surface Roughness

Normalized Shear Stress

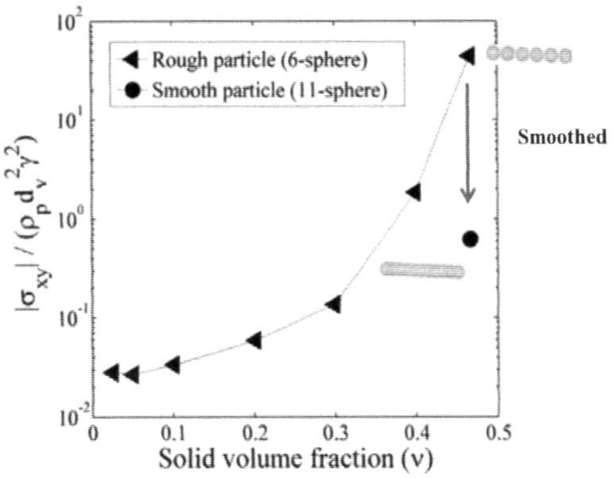

(Fiber Aspect Ratio AR = 6, E_b = 2.6 e4 Pa, β_c = 1.63e-2, β_b = 0, μ = 0.5)

In dense flows of flexible fibers, the smooth surfaces allow the fibers to slide past their neighbors more easily, leading to the reduction in **particle deformation** and **stress**

UF UNIVERSITY of FLORIDA
The Foundation for The Gator Nation

26

Shearing of Wet, Elongated, Flexible Particles

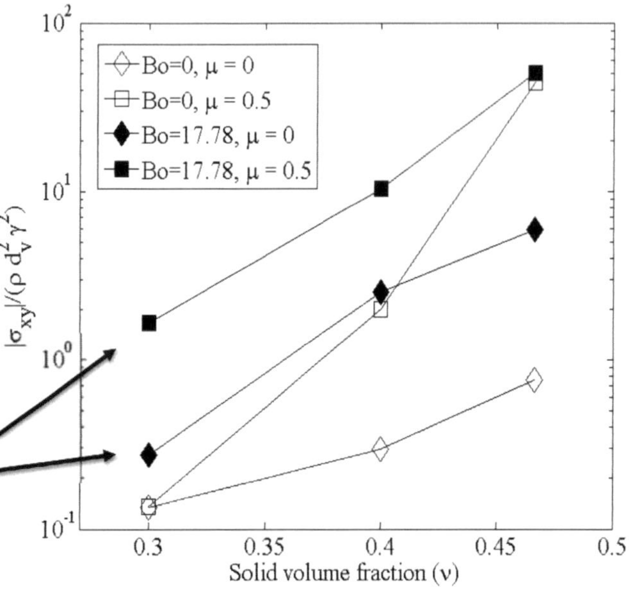

Liquid Bridging

Effect of Particle Flexibility

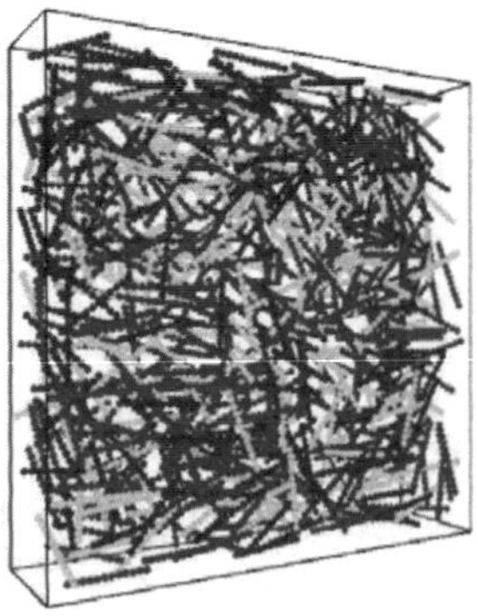

Bond Young's Modulus Eb = 1.0 e4 Pa
Coordination number = 5.2

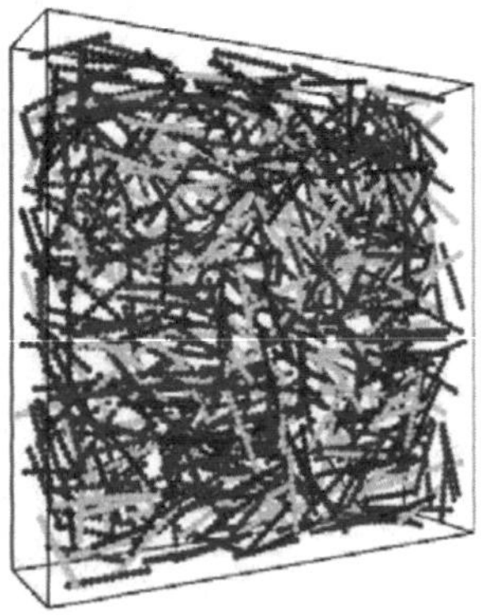

Bond Young's Modulus Eb = 1.0 e2 Pa
Coordination number = 8.3

Increases in Particle Flexibility Increases Particle Deformation as well as Packing Density of Agglomerates

Aspect Ratio = 12, Solid fraction = 0.05, α = 4e-3 N/m, γ = 100/s, Bo = 17.78

Effect of Particle Aspect Ratio

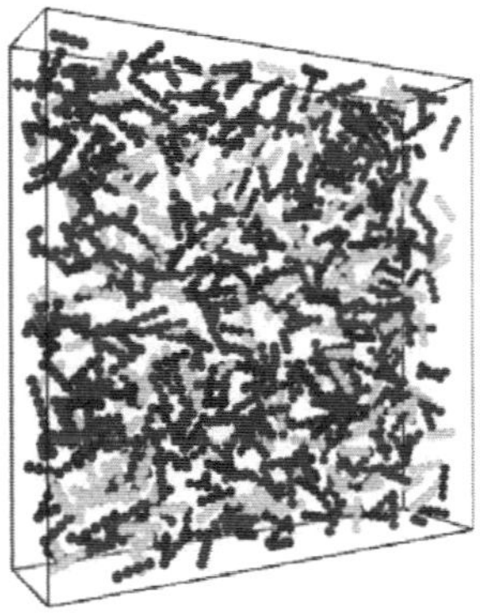

Aspect Ratio = 4

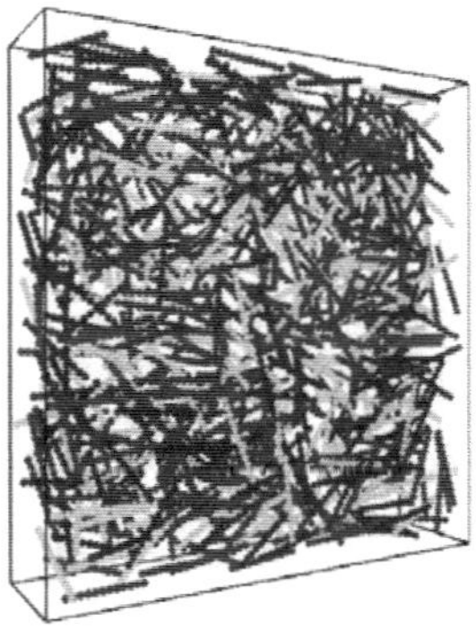

Aspect Ratio = 12

With a larger particle aspect ratio, larger clumps are obtained and the fibers tend to be aligned in the clumps

Solid fraction = 0.05, α = 4e-3 N/m, γ = 100/s, Bo = 17.78

UF | UNIVERSITY of FLORIDA
The Foundation for The Gator Nation

Publications To Date

- Y. Guo, C. Wassgren, W. Ketterhagen, B. Hancock, B. James and J. Curtis, "A Numerical Study of Granular Shear Flows of Rod-like Particles Using the Discrete Element Method", *Journal of Fluid Mechanics*, 713, 1-26 (2012)

- Y. Guo, C. Wassgren, B. Hancock, W. Ketterhagen, and J. Curtis, "Granular Shear Flows of Flat Disks and Elongated Rods without and with Friction," *Physics of Fluids*, 25, Article 06334, 25 pages (2013)

- Y. Guo and J. Curtis, "Discrete Element Method Simulations for Complex Granular Flows" (Invited), *Annual Review of Fluid Mechanics*, in press (2014)

UF | UNIVERSITY of FLORIDA
The Foundation for The Gator Nation

Angle of Repose and Shear Cell Testing

Dry steel cylinders

Wet steel cylinders

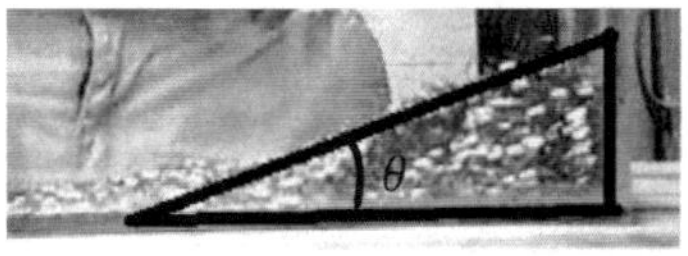

$\theta = 19.2°$

$\theta = 32.3°$

UF | UNIVERSITY of FLORIDA
The Foundation for The Gator Nation

Flexible Fiber Model

- ❑ **Contact damping is introduced to account for energy dissipation in contact**
  - **Normal contact damping force:** $\boldsymbol{F}_{nd}^{c} = 2\beta_c\sqrt{m^* K_n^c}\,\boldsymbol{v}_n^r$
  - **Tangential contact damping force:** $\boldsymbol{F}_{td}^{c} = 2\beta_c\sqrt{m^* K_t^c}\,\boldsymbol{v}_t^r$
- ❑ **Bond damping is introduced to account for energy dissipation in fiber vibration**
  - **Bond normal damping force:** $\boldsymbol{F}_{nd}^{b} = 2\beta_b\sqrt{m^* K_n^b}\,\boldsymbol{v}_n^r$
  - **Bond tangential damping force:** $\boldsymbol{F}_{td}^{b} = 2\beta_b\sqrt{m^* K_t^b}\,\boldsymbol{v}_t^r$
  - **Bond bending damping moment:** $\boldsymbol{M}_{td}^{b} = 2\beta_b\sqrt{J^* K_{ben}^b}\,\boldsymbol{\omega}_t^r$
  - **Bond twisting damping moment:** $\boldsymbol{M}_{nd}^{b} = 2\beta_b\sqrt{J^* K_{twi}^b}\,\boldsymbol{\omega}_n^r$

$m^* = m_1 m_2/(m_1 + m_2)$ and $J^* = J_1 J_2/(J_1 + J_2)$
m_1, m_2, J_1, J_2 → Mass and moment of inertia of two contacting/bonding spheres
$K_n^b, K_t^b, K_{ben}^b, K_{twi}^b$ → Bond normal, tangential, bending, and twisting stiffnesses
β_c, β_b → Contact damping coefficient and bond damping coefficient
v_n^r, v_t^r → Relative normal and tangential velocities between two contacting/bonded spheres
ω_n^r, ω_t^r → Normal and tangential components of relative angular velocity between two bonded spheres

32

UF | UNIVERSITY of FLORIDA
The Foundation for The Gator Nation

Flexible Fiber Model

- **Translational and rotational motions of spheres in each fiber are governed by Newton's second law of motion**

$$m_s \frac{d^2 \boldsymbol{x}_s}{dt^2} = \boldsymbol{F}^b + \boldsymbol{F}^c$$

$$J_s \frac{d^2 \boldsymbol{\theta}_s}{dt^2} = \boldsymbol{M}^b + \boldsymbol{M}^c$$

m_s , J_s	→	Mass and moment of inertia of a sphere
$\boldsymbol{x}_s$, $\boldsymbol{\theta}_s$	→	Translational and angular displacements of a sphere
$\boldsymbol{F}^b$	→	Bond forces
$\boldsymbol{F}^c$	→	Contact forces due to the contacts with the spheres from the other fibers
$\boldsymbol{M}^b$	→	Bond bending moments, twisting moments, and the moments due to bond shear forces
$\boldsymbol{M}^c$	→	Moments due to contact forces

33

UF | UNIVERSITY of FLORIDA
The Foundation for The Gator Nation

Wet Granular Flows and Structures

Kudrolli, A., Clark University

Originally presented on: 5/1/2014 10:30:00 - 10:55:00

We will examine the properties of wet granular matter where the interplay of capillary and viscous forces and inertia can be important to their observed properties. We will first review work on our understanding of the impact of cohesive partially wetting liquid on the stability and flow properties of wet granular matter and particle segregation. Then, we will examine how these properties can be exploited to self-assemble a wide variety of intricately designed slender pillar and arch structures by simply pouring a granular suspension onto a liquid absorbing surface. We will discuss its potential for rapid manufacturing techniques such as 3D printing. Then, we will examine the interaction of liquid flow with sedimented granular beds leading to onset of motion and development of heterogeneity. Implications to boreholes and other geotechnical problems will be discussed.

Electrostatic charging of flowing granular materials

Daniel Lacks

Department of Chemical Engineering

Case Western Reserve University

CASE SCHOOL OF ENGINEERING
CASE WESTERN RESERVE UNIVERSITY

Flavor particles on potato chips

Yousuf and Barringer (2007)

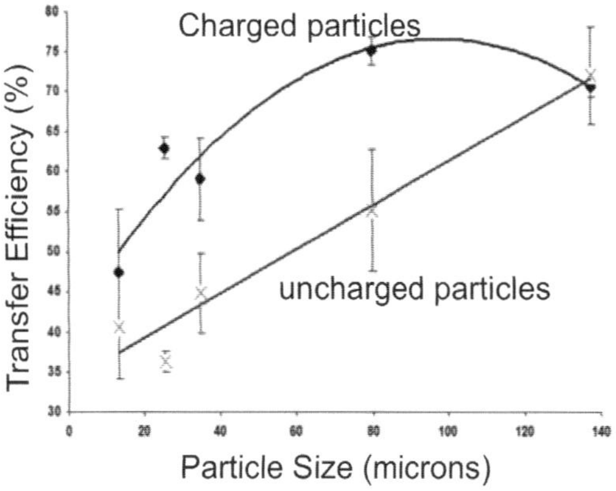

CASE SCHOOL OF ENGINEERING
CASE WESTERN RESERVE UNIVERSITY

Fluidized bed reactors

Unwanted consequence: particles rub against each other and become electrostatically charged

CASE SCHOOL
OF ENGINEERING
CASE WESTERN RESERVE
UNIVERSITY

Industrial powders

Inculet, Castle, Aartsen, *Chem. Eng. Sci.* (2006)

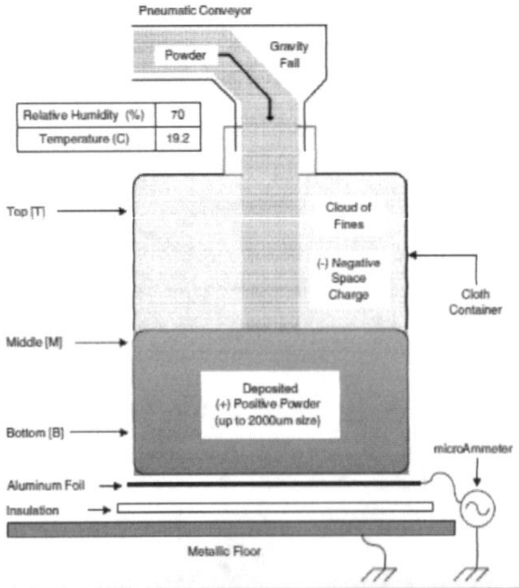

CASE SCHOOL
OF ENGINEERING
CASE WESTERN RESERVE
UNIVERSITY

Industrial powders

Inculet, Castle, Aartsen, *Chem. Eng. Sci.* (2006)

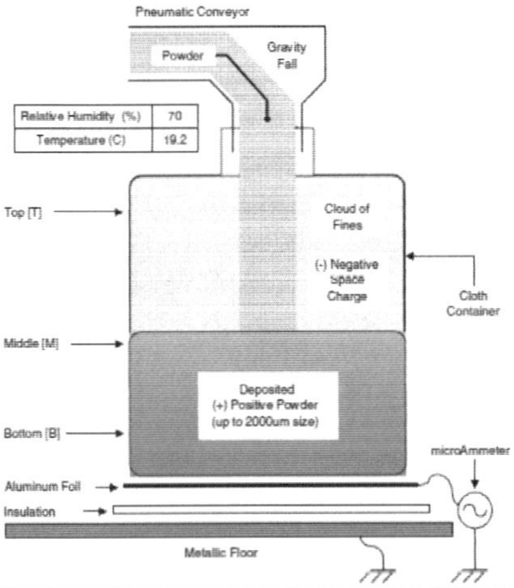

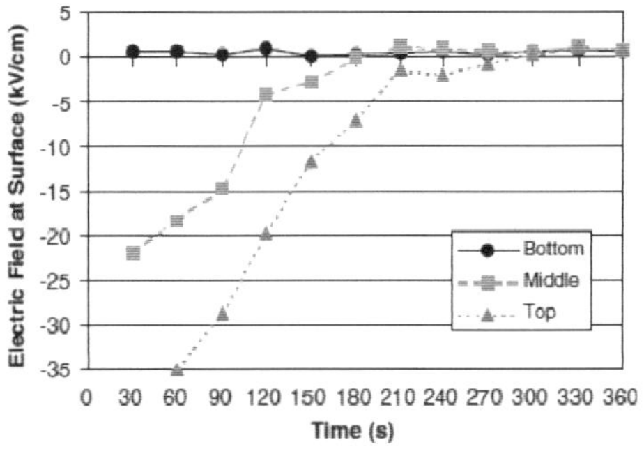

Negative electric field implies smaller particles charge negatively

CASE SCHOOL OF ENGINEERING
CASE WESTERN RESERVE UNIVERSITY

Iceland, 2010

CASE SCHOOL OF ENGINEERING
CASE WESTERN RESERVE UNIVERSITY

Volcanic plumes

Miura et al, *Bull Volcanol.* (2002)

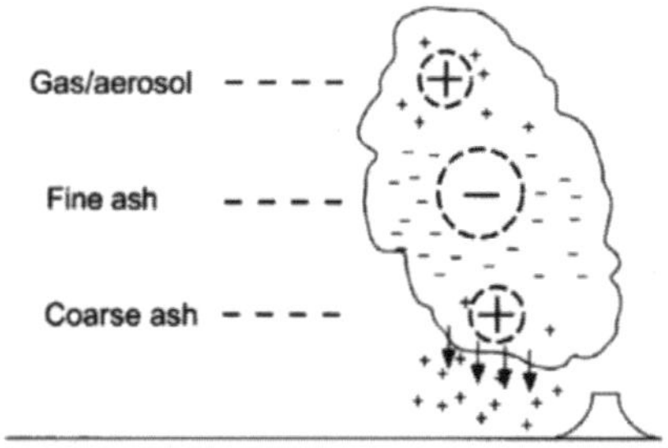

"Volcanic plumes are generally composed of three parts: an upper part with positively charged gas and aerosol, a middle part with **negatively charged fine ash particles**, and a lower part with **positively charged coarse ash particles**."

CASE SCHOOL OF ENGINEERING
CASE WESTERN RESERVE UNIVERSITY

Niger, 2010

CASE SCHOOL OF ENGINEERING
CASE WESTERN RESERVE UNIVERSITY

Dust devils

Farrell et al, *J. Geophys. Res.* (2004)

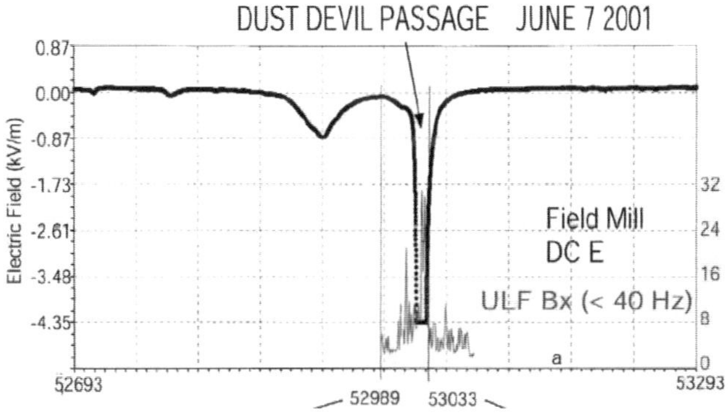

Smaller particles lofted higher above surface

Negative electric field implies smaller particles charge negatively

CASE SCHOOL OF ENGINEERING
CASE WESTERN RESERVE UNIVERSITY

Earliest text mentioning electrostatics

Plato, *Timaeus* (ca 360 BC)

"Moreover, as to the flowing of water, the fall of the thunderbolt, and the marvels that are observed about the attraction of amber and the Heraclean stones, in none of these cases is there any attraction; but he who investigates rightly, will find that such wonderful phenomena are attributable to the combination of certain conditions the non-existence of a vacuum"

Plato (left) carrying the *Timaeus* in painting by Raphael in the Sistine chapel

CASE SCHOOL OF ENGINEERING
CASE WESTERN RESERVE UNIVERSITY

Scientific progress since 360 BC

In *Timaeus*, Plato discusses:

- astronomy

- materials

- magnetism

- electrostatic charging

CASE SCHOOL OF ENGINEERING
CASE WESTERN RESERVE UNIVERSITY

Scientific progress since 360 BC

In *Timaeus*, Plato discusses:

- astronomy

2014: send people to the moon and spaceships to other planets

- materials

- magnetism

- electrostatic charging

CASE SCHOOL OF ENGINEERING
CASE WESTERN RESERVE UNIVERSITY

Scientific progress since 360 BC

In *Timaeus*, Plato discusses:

- astronomy
2014: send people to the moon and spaceships to other planets

- materials
2014: create new materials not found in nature, with properties tailored to our needs

- magnetism

- electrostatic charging

CASE SCHOOL
OF ENGINEERING
CASE WESTERN RESERVE
UNIVERSITY

Scientific progress since 360 BC

In *Timaeus*, Plato discusses:

- astronomy
2014: send people to the moon and spaceships to other planets

- materials
2014: create new materials not found in nature, with properties tailored to our needs

- magnetism
2014: MRI to produce three-dimensional images to diagnose diseases

- electrostatic charging

CASE SCHOOL
OF ENGINEERING
CASE WESTERN RESERVE
UNIVERSITY

Scientific progress since 360 BC

In *Timaeus*, Plato discusses:

- astronomy
2014: send people to the moon and spaceships to other planets

- materials
2014: create new materials not found in nature, with properties tailored to our needs

- magnetism
2014: MRI to produce three-dimensional images to diagnose diseases

- electrostatic charging
2014: Little progress since 360 BC

CASE SCHOOL OF ENGINEERING
CASE WESTERN RESERVE UNIVERSITY

Can the direction of charging be predicted?

'Triboelectric series'

	Wilcke 1759
+	Glass
	Wool
	Quills
	Wood
	Paper
	Ground glass
	Pb
	Sulphur
-	Metals

CASE SCHOOL OF ENGINEERING
CASE WESTERN RESERVE UNIVERSITY

Can the direction of charging be predicted?

the answer is … *sometimes, and with no theoretical understanding*

'Triboelectric series'

Wilcke 1759	Faraday 1840	Jamin, Bouty 1891	Shaw *Pr. Roy. Soc.* 1917	Hersch et al. *Text. Res. J.* 1955	Henneker *Nature* 1962
Glass	Cat's fur	Cat's fur	Glass	Wool	Nylon
Wool	Wool	Glass	Wool	Nylon	Wool
Quills	Quills	Wool	Cat's fur	**Cotton**	**Silk**
Wood	Flint glass	Feathers	**Silk**	**Silk**	**Cotton**
Paper	**Cotton**	Wood	Paper	PVC	Polyethylene
Ground glass	Linen	Paper	**Cotton**	Polyethylene	PVC
Pb	**Silk**	Silk	Wood, Fe		
Sulphur	Wood	Resin	Ground glass		
Metals	Fe,Cu,Ag,Pb	Ground glass	Cu, Ag		
	Sulphur		Sulfur		

(+ at top, − at bottom)

CASE SCHOOL OF ENGINEERING
CASE WESTERN RESERVE UNIVERSITY

Can the direction of charging be predicted?

Baytekin et al., *Angew. Chem.* (2012)

Roll PS beads in a PTFE dish

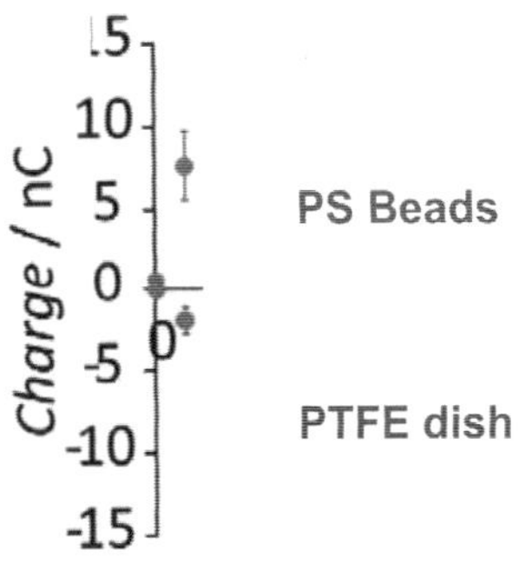

Henneker (Nature,1962)
Nylon
Wool
Silk
Cotton
Polystyrene
Polyethylene
PVC
PTFE

(+ at top, − at bottom)

CASE SCHOOL OF ENGINEERING
CASE WESTERN RESERVE UNIVERSITY

Can the direction of charging be predicted?

Baytekin et al., *Angew. Chem.* (2012)

Roll PS beads in a PTFE dish

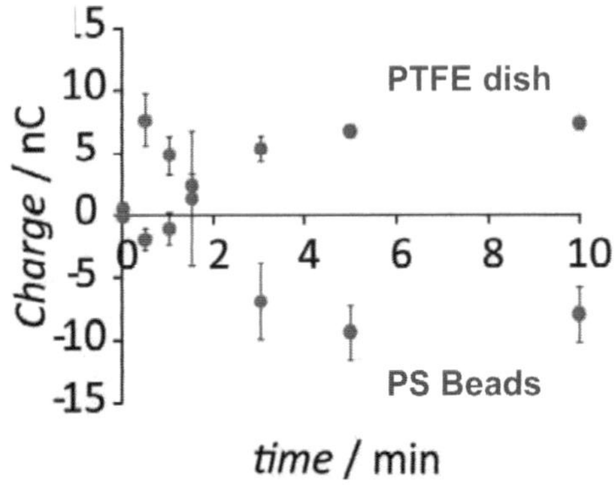

Henneker (Nature,1962)

+

Nylon
Wool
Silk
Cotton
Polystyrene
Polyethylene
PVC
PTFE

-

CASE SCHOOL OF ENGINEERING
CASE WESTERN RESERVE UNIVERSITY

What is being transferred?

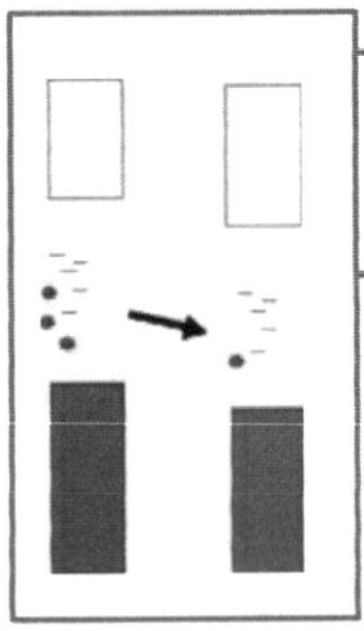

Liu and Bard, *Nature Materials* (2008)
"electrostatic charges on Teflon produced by rubbing with Lucite were directly identified as **electrons** rather than ions"

CASE SCHOOL OF ENGINEERING
CASE WESTERN RESERVE UNIVERSITY

What is being transferred?

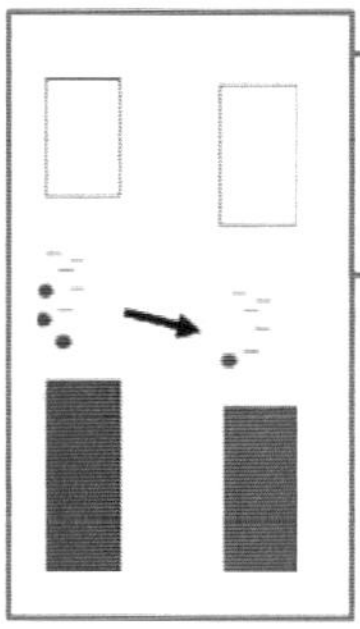

Liu and Bard, *Nature Materials* (2008)
"electrostatic charges on Teflon produced by rubbing with Lucite were directly identified as **electrons** rather than ions"

McCarty and Whitesides, *Angew. Chem.* (2008)
"this **ion-transfer** mechanism may also explain the ubiquitous contact electrification ("static electricity") of materials"

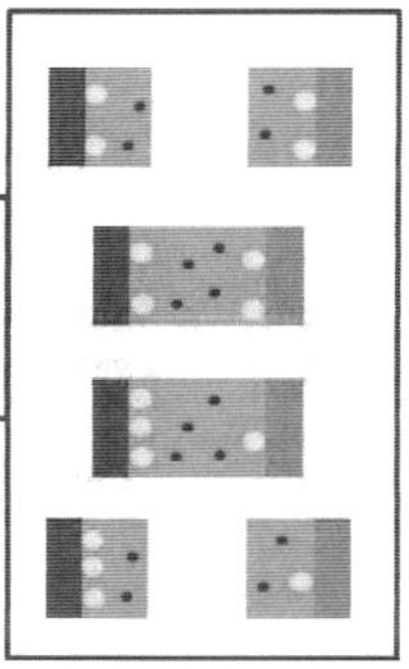

CASE SCHOOL OF ENGINEERING
CASE WESTERN RESERVE UNIVERSITY

What is being transferred?

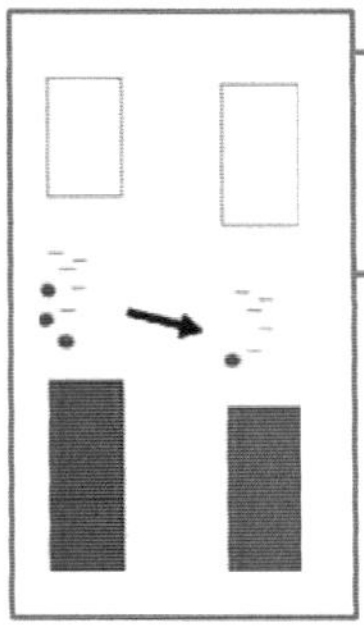

Liu and Bard, *Nature Materials* (2008)
"electrostatic charges on Teflon produced by rubbing with Lucite were directly identified as **electrons** rather than ions"

McCarty and Whitesides, *Angew. Chem.* (2008)
"this **ion-transfer** mechanism may also explain the ubiquitous contact electrification ("static electricity") of materials"

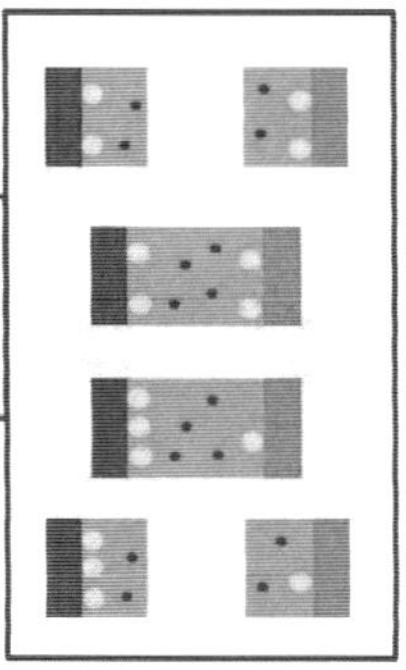

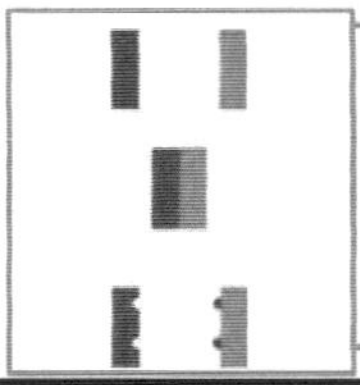

CBS News.com, June 29, 2011
"New research published in the journal Science shows that static electricity is caused by more than just the exchange of individual electrons, and instead involves the transfer of bigger (yet still tiny) **clumps of material**."
[Baytekin et. al, *Science* (2011)]

CASE SCHOOL OF ENGINEERING
CASE WESTERN RESERVE UNIVERSITY

Collaborators

Prof. Mohan Sankaran
Dr. Keith Forward

CASE SCHOOL OF ENGINEERING
CASE WESTERN RESERVE UNIVERSITY

Collaborators

Dr. Mamadou Sow

CASE SCHOOL OF ENGINEERING
CASE WESTERN RESERVE UNIVERSITY

Collaborators

Dr. Mihai Bilici

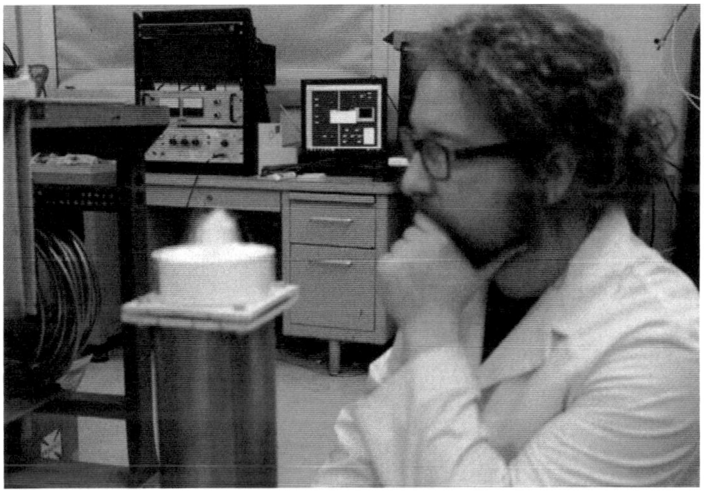

CASE SCHOOL OF ENGINEERING
CASE WESTERN RESERVE UNIVERSITY

Effects of material strain

We use a very simple experiment
... but the underlying physics is complex

- Rub balloon on Teflon® PTFE
- Try both inflated and deflated balloons
- Measure surface potential on Teflon® PTFE after rubbing with balloon

CASE SCHOOL OF ENGINEERING
CASE WESTERN RESERVE UNIVERSITY

Home-made balloon

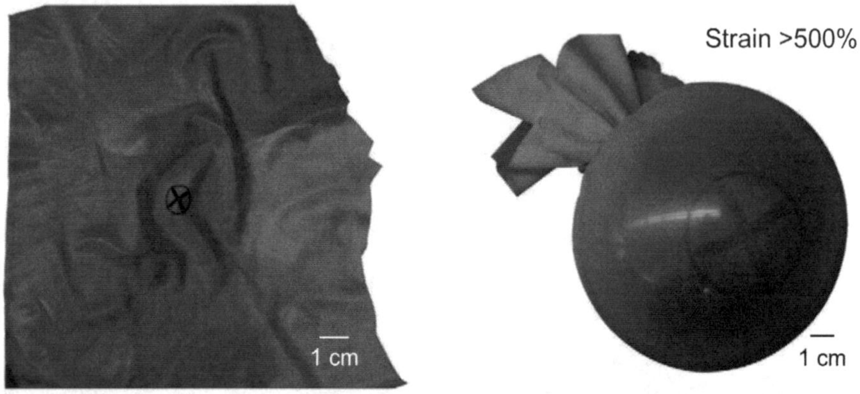

CASE SCHOOL OF ENGINEERING
CASE WESTERN RESERVE UNIVERSITY

Start with neutral Teflon® PTFE

Sow et al., *Angew Chem.* (2012)

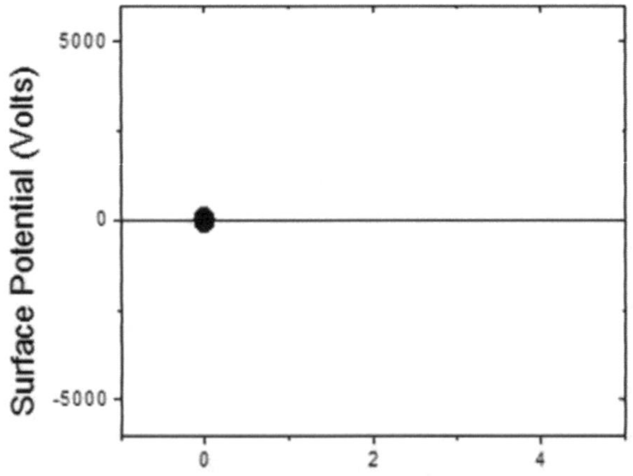

CASE SCHOOL OF ENGINEERING
CASE WESTERN RESERVE UNIVERSITY

...then rub with inflated balloon

Sow et al., *Angew Chem.* (2012)

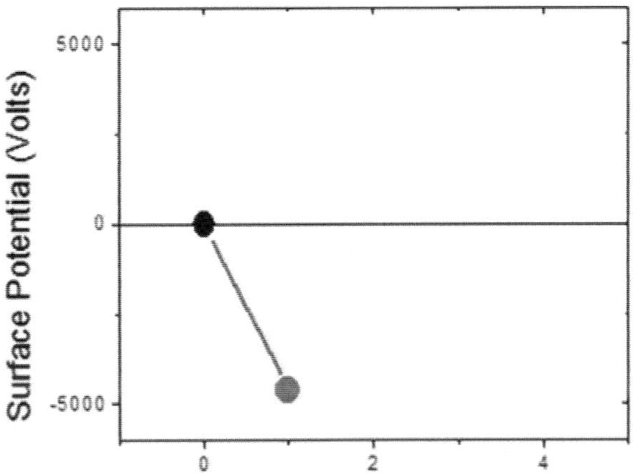

CASE SCHOOL OF ENGINEERING
CASE WESTERN RESERVE UNIVERSITY

...then rub with deflated balloon

Sow et al., *Angew Chem.* (2012)

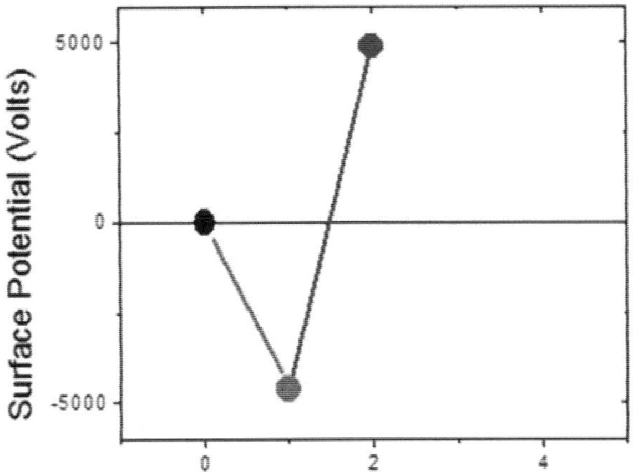

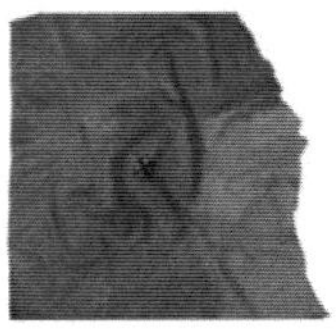

CASE SCHOOL OF ENGINEERING
CASE WESTERN RESERVE UNIVERSITY

...then rub with inflated balloon

Sow et al., *Angew Chem.* (2012)

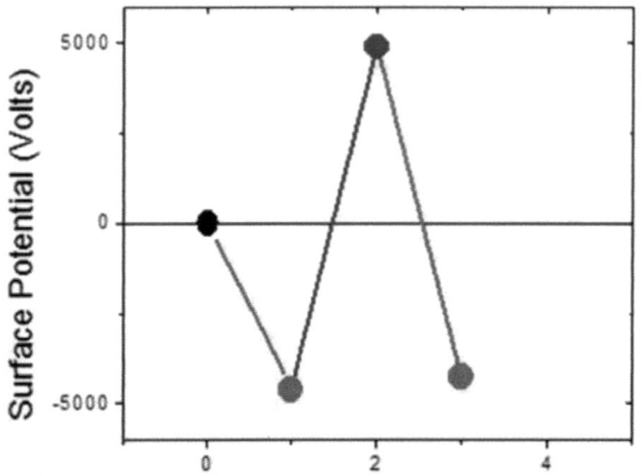

CASE SCHOOL
OF ENGINEERING
CASE WESTERN RESERVE
UNIVERSITY

...then rub with deflated balloon

Sow et al., *Angew Chem.* (2012)

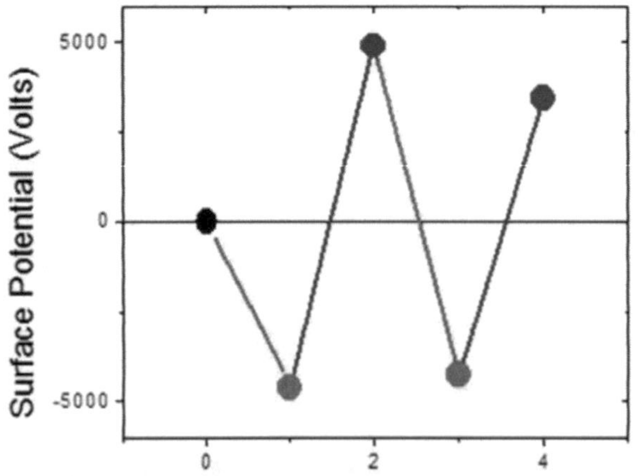

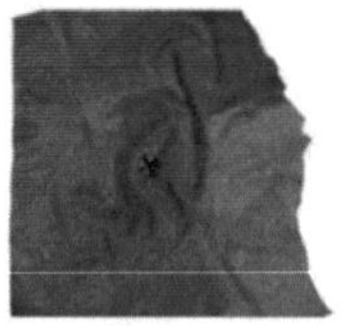

CASE SCHOOL
OF ENGINEERING
CASE WESTERN RESERVE
UNIVERSITY

Results are totally reproducible*

* when samples carefully cleaned — Sow et al., *Angew Chem.* (2012)

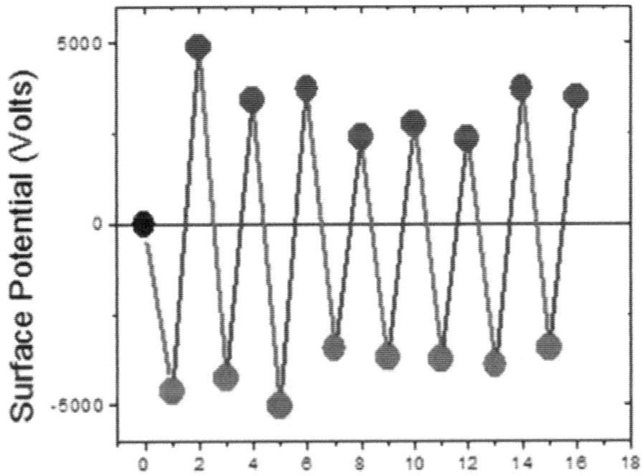

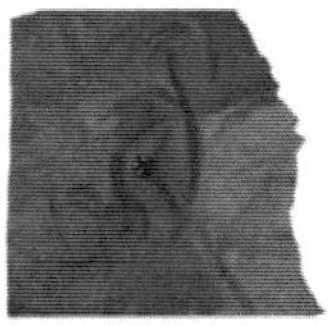

CASE SCHOOL OF ENGINEERING
CASE WESTERN RESERVE UNIVERSITY

Nanoscale strains in materials

Forward et al., *Langmuir* (2007)

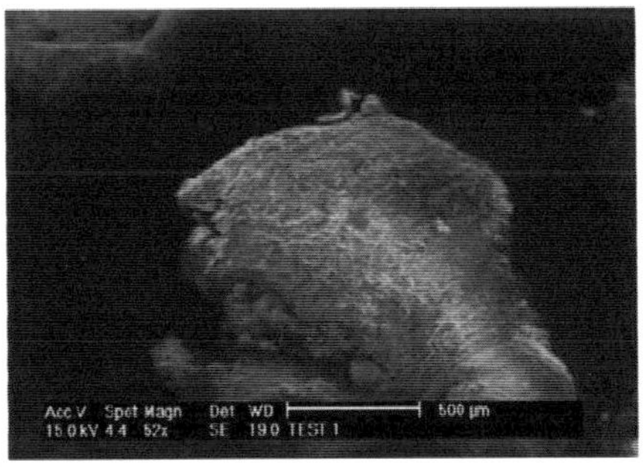

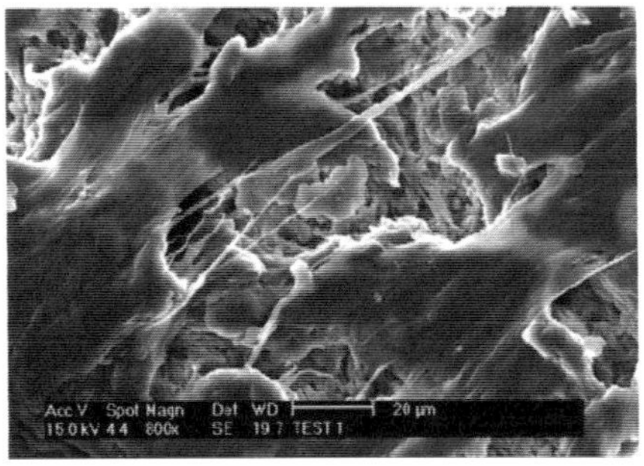

-Even if material not macroscopically strained, strains exist at nanoscale
-Our results suggest the nanoscale strains will affect charging
-Nanoscale strains dependent on material history (thermal, mechanical)
→ **Nanoscale strains can lead to unpredictability of charging**

CASE SCHOOL OF ENGINEERING
CASE WESTERN RESERVE UNIVERSITY

Electrostatic charging of granular systems

Challenges:

- Electrostataic charging occurs whenever two surfaces contact – makes it difficult to study!
- Charging in granular systems can occur in 2 ways: particle – wall or particle – particle
- Bipolar charging occurs from particle-particle interactions; net charge not so meaningful

We developed a methodology to:

- Disentangle the particle - wall and particle - particle effects
- Measure distribution of charge in particle systems (not just net charge)

CASE SCHOOL OF ENGINEERING
CASE WESTERN RESERVE UNIVERSITY

particle-particle charging

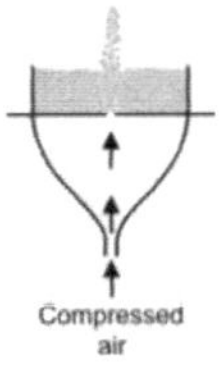

CASE SCHOOL OF ENGINEERING
CASE WESTERN RESERVE UNIVERSITY

particle-particle charging

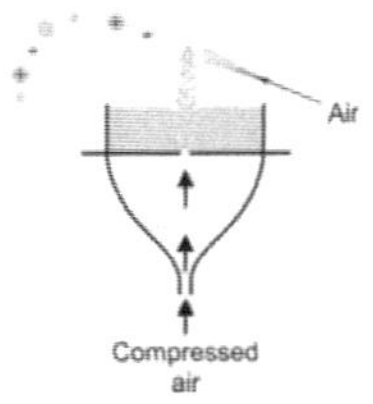

CASE SCHOOL
OF ENGINEERING
CASE WESTERN RESERVE
UNIVERSITY

particle-particle charging

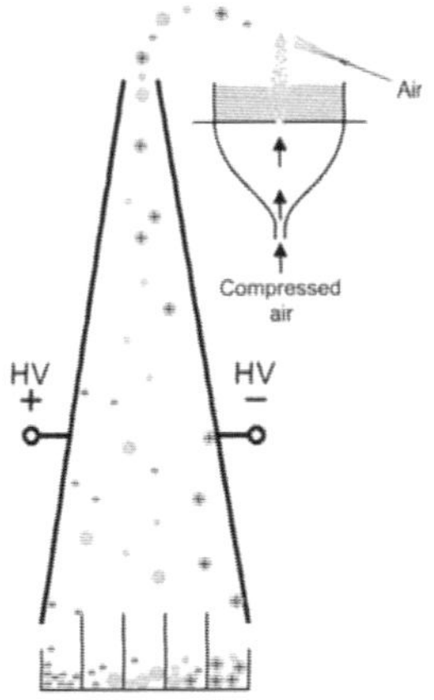

5 containers (Faraday cups)

CASE SCHOOL
OF ENGINEERING
CASE WESTERN RESERVE
UNIVERSITY

particle-particle charging

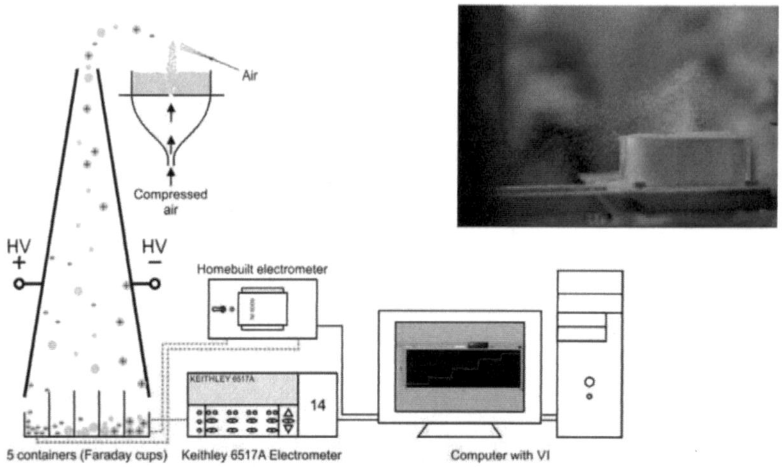

CASE SCHOOL
OF ENGINEERING
CASE WESTERN RESERVE
UNIVERSITY

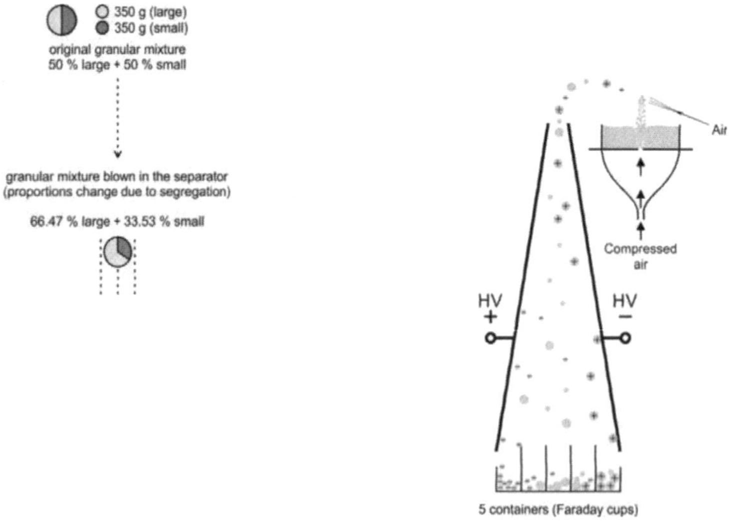

CASE SCHOOL
OF ENGINEERING
CASE WESTERN RESERVE
UNIVERSITY

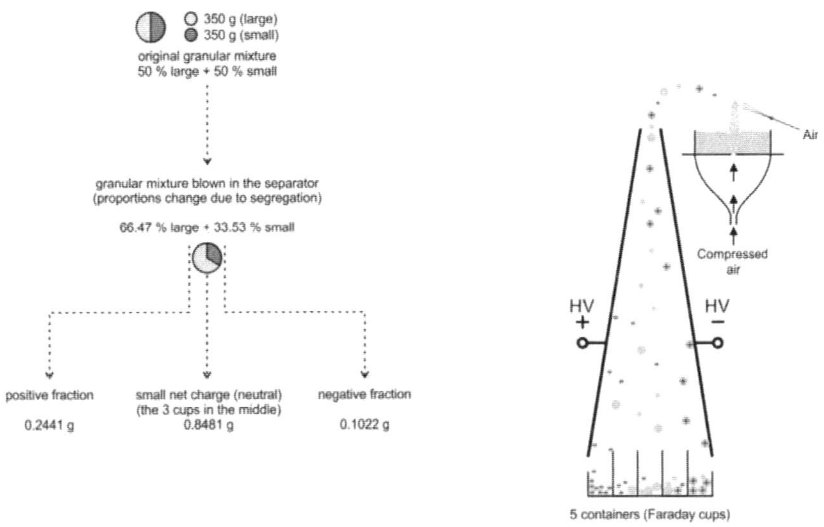

CASE SCHOOL
OF ENGINEERING
CASE WESTERN RESERVE
UNIVERSITY

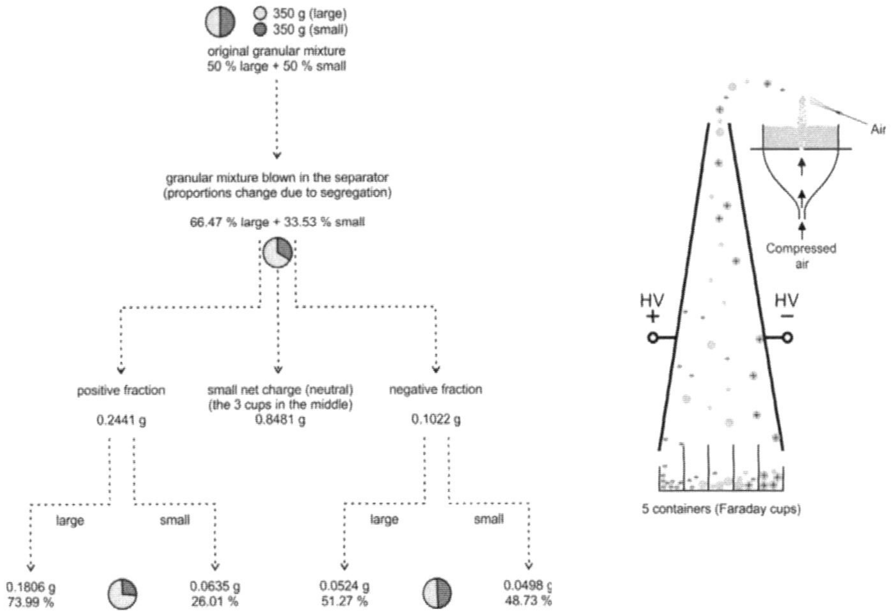

CASE SCHOOL
OF ENGINEERING
CASE WESTERN RESERVE
UNIVERSITY

Results

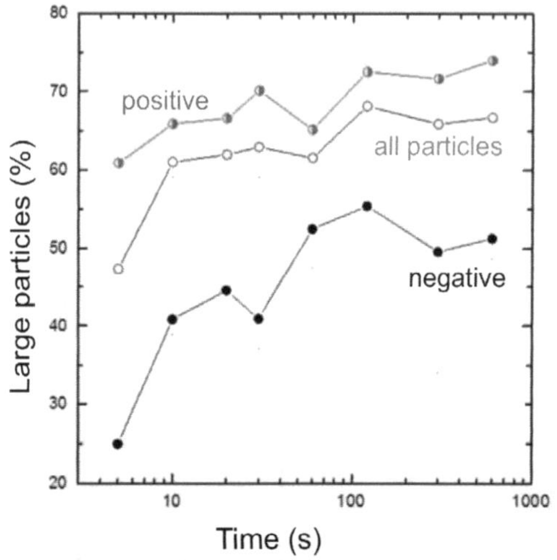

CASE SCHOOL OF ENGINEERING
CASE WESTERN RESERVE UNIVERSITY

Experiments at University of Chicago

Waitukaitis and Jaeger, in preparation

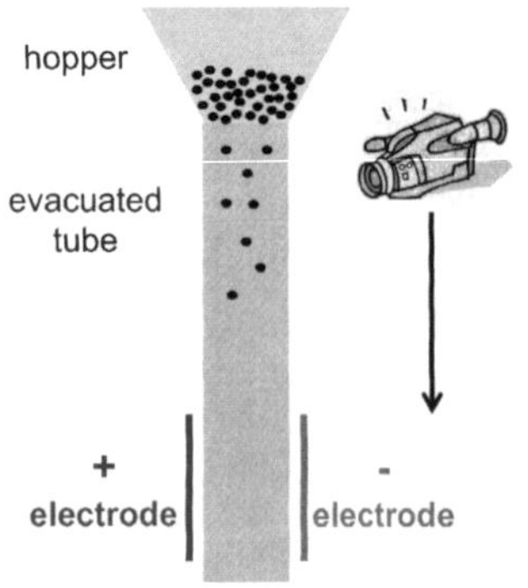

High-speed camera falls at same velocity as falling particles, and films particle trajectories as they move through applied electric field

CASE SCHOOL OF ENGINEERING
CASE WESTERN RESERVE UNIVERSITY

Experiments at University of Chicago

Waitukaitis and Jaeger, in preparation

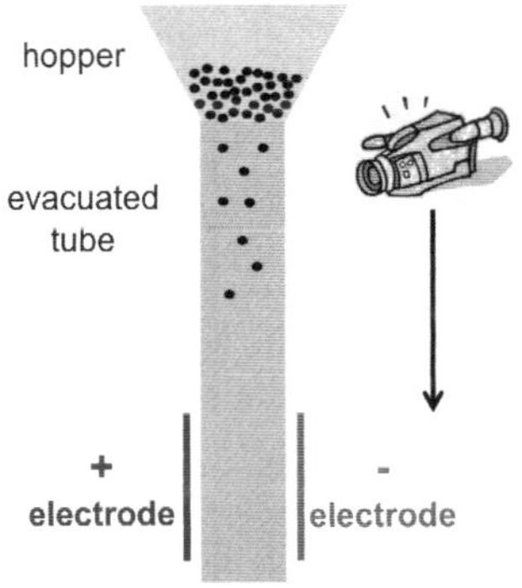

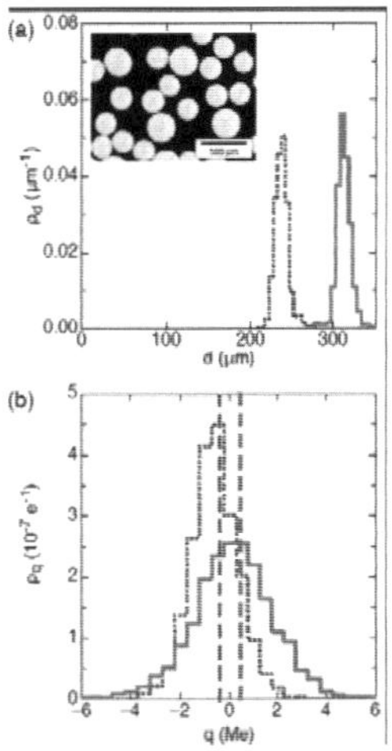

CASE SCHOOL OF ENGINEERING
CASE WESTERN RESERVE UNIVERSITY

Why particle-size dependent charging?

Lacks, Duff, Kumar, *PRL* (2008)

We adapt non-equilibrium model for asymmetric rubbing (Lowell and Truscott, J. Phys. D, 1986) to particle systems

'Money-exchange game'
- bring 1% of savings in $1 bills, in **left** pocket
- everyone blindfolded
- when collide with someone, give them $1 from **left** pocket
- put money you get into **right** pocket

CASE SCHOOL OF ENGINEERING
CASE WESTERN RESERVE UNIVERSITY

Particle dynamics simulations

Lacks, Duff, Kumar, *PRL* (2008)

Simulation methodology

- 864 particles of different sizes
- Newtonian dynamics
- Transfer electrons at each collision according to trapped electron model
- Run simulation for 10 million collisions

Results for different densities

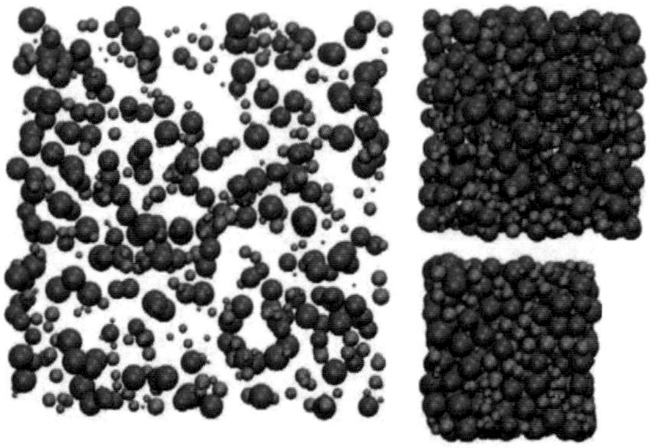

Blue: positive **Red: negative**

CASE SCHOOL OF ENGINEERING
CASE WESTERN RESERVE UNIVERSITY

Particle dynamics simulations

Lacks, Duff, Kumar, *PRL* (2008)

Simulation methodology

- 864 particles of different sizes
- Newtonian dynamics
- Transfer electrons at each collision according to trapped electron model
- Run simulation for 10 million collisions

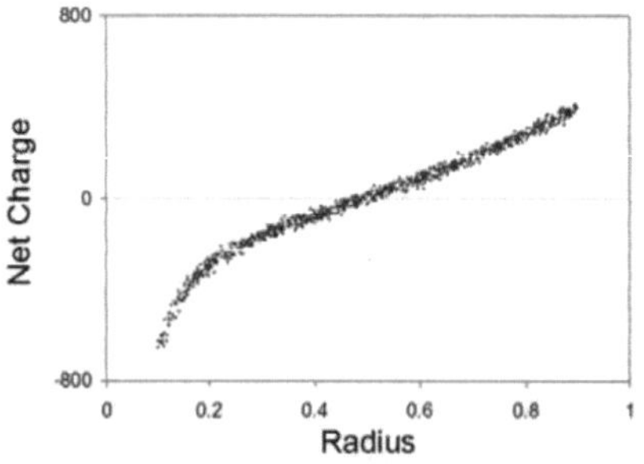

CASE SCHOOL OF ENGINEERING
CASE WESTERN RESERVE UNIVERSITY

Conclusion

- Measuring the electrostatic charging in granular systems not so straighforward!
- Interesting and subtle phenomena, such as particle size dependence of charging

Relevant papers:
- Lacks, Duff, Kumar, *PRL* (2008)
- Forward, Lacks, Sankaran, *PRL* (2009)
- Sow, Widenor, Akande, Robinson, Sankaran, Lacks, *J. Brazil Chem. Soc.* (2012)
- Bilici, Toth, Lacks, Sankaran, in preparation

Perspectives articles, reviews:
- Lacks, *Nature Physics* (2010)
- Lacks and Sankaran, *J. Phys. D* (2011)
- Lacks, *Angew. Chem.* (2012)

Funding from NSF:
DMR-0705191,1206480
CBET-0852773, 1235908

CASE SCHOOL
OF ENGINEERING
CASE WESTERN RESERVE
UNIVERSITY

Particles Interactions Resulting in Electrostatic Charge Generation in Gas-Solid Fluidized Beds

Mehrani, P., University of Ottawa

Originally presented on: 5/1/2014 11:30:00 - 11:55:00

In any gas-solid processing system, generation of electrostatic charges is inevitable due to particles interactions (i.e., continuous contacts between particles, and the particles and the containing vessel). Electrostatic charging could in turn result in problems of inter-particle cohesion or vessel wall adhesion. The extent of the net exchange of electrostatic charge between any surfaces depends on various parameters including the materials electrical properties, surface chemistry, environmental conditions and many more. This talk will focus on electrostatic charge generation in gas-solid fluidized beds and issues of particle agglomeration and reactor fouling in some commercial operations, and presents common measurement techniques applied for fluidizing particles electrostatic charge characterization as well as a brief summary of parameters that could affect the degree of particles charging.

Scale-Up of Agitated Drying: Effect of Shear Stress and Hydrostatic Pressure on API Powder Properties

Remy, B., Bristol-Myers Squibb

Originally presented on: 5/1/2014 13:00:00 - 13:25:00

Roughly 80% of commercial pharmaceutical products utilize oral solid dosage formulations. In these formulations, the Active Pharmaceutical Ingredient (API) particle size can influence stability, bioavailability, dissolution rate and content uniformity of the drug product. For this reason, the pharmaceutical industry devotes significant effort to developing fundamental understanding of the processes that govern API particle size. Much of that effort has historically focused on crystallization development, whereas fundamental understanding of particle agglomeration and attrition in the subsequent agitated drying step is less well established. One reason is that traditional laboratory drying equipment has not successfully reproduced the degree of agglomeration or attrition observed at scale. This work characterizes the propensity for particle attrition to occur at different stages of the drying process by employing a laboratory agitated dryer that has been modified to measure impeller torque and reproduce the range of hydrostatic pressures observed during scale-up. A workflow for assessing risk of API attrition at manufacturing scale is described.

Open Questions in Modeling Collision, Coalescence and Breakage Processes

Diemer, R. B., DuPont Company

Originally presented on: 5/1/2014 13:30:00 - 13:55:00

Population balance models are potentially useful for design and scaleup of particle processing operations, but much is missing that would enable broader application of these approaches to real problems. This paper will describe some of the gaps that need to be closed in order to better connect experiments, models and equipment design.

The Importance of Fluid-Particle Interactions for Fluidized Systems Modeling

Blaser, P., CPFD Software, LLC

Originally presented on: 5/1/2014 14:00:00 - 14:25:00

This presentation discusses the importance of capturing fluid-particle interactions in computational models of fluidized systems. While direct particle to particle interactions are certainly important and computational modeling in general would benefit from an improved understanding of these interactions, a primary interaction mechanism influencing fluidized systems is the interaction between the particles and fluid. This presentation discusses the importance of fluid-particle coupling via aerodynamic/hydrodynamic drag, how particle-fluid interactions impact particle to particle interactions, validation of modeling for fluidized industrial systems, the effect of particle size distribution including fines, limitations of existing models, and how current research trends can enable expanded industrial application of modeling technology impacting efficient and rapid commercialization and scale-up of new and improved particle technologies.

AUTHOR INDEX